Illustration: Magali Lefrançois.

David Thompson, 1770-1857.

David Thompson

Tom Shardlow

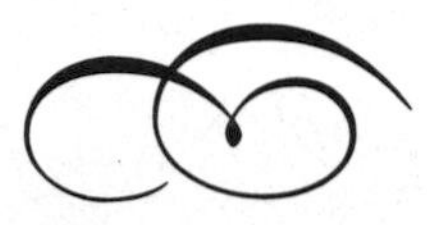

Tom Shardlow is a research biologist who has more than thirty articles published in international scientific and technical journals. His diverse writing ranges from poetry and award-winning short fiction published in literary magazines to book reviews for *Quill and Quire* and *Canadian Geographic*. His freelance work, which includes creative drawings and photographs, often appears as feature articles in magazines and newspapers. Currently, he is working on two non-fiction books, one for children, the other a popular history of biology, and he is completing a science-fiction novel. Tom Shardlow lives with his family on Vancouver Island. Comments on his books may be sent to tomshardlow@shaw.ca

Editorial correspondence:
Rhonda Bailey, Editorial Director
XYZ Publishing
P.O. Box 250
Lantzville, BC
V0R 2H0
E-mail: xyzed@shaw.ca

In the same collection

Ven Begamudré, *Isaac Brock: Larger Than Life.*
Lynne Bowen, *Robert Dunsmuir: Laird of the Mines.*
Kate Braid, *Emily Carr: Rebel Artist.*
Kathryn Bridge, *Phyllis Munday: Mountaineer.*
William Chalmers, *George Mercer Dawson: Geologist, Scientist, Explorer.*
Anne Cimon, *Susanna Moodie: Pioneer Author.*
Deborah Cowley, *Lucille Teasdale: Doctor of Courage.*
Gary Evans, *John Grierson: Trailblazer of Documentary Film.*
Judith Fitzgerald, *Marshall McLuhan: Wise Guy.*
lian goodall, *William Lyon Mackenzie King: Dreams and Shadows.*
Stephen Eaton Hume, *Frederick Banting: Hero, Healer, Artist.*
Naïm Kattan, *A.M. Klein: Poet and Prophet.*
Betty Keller, *Pauline Johnson: First Aboriginal Voice of Canada.*
Heather Kirk, *Mazo de la Roche: Rich and Famous Writer.*
Michelle Labrèche-Larouche, *Emma Albani: International Star.*
Wayne Larsen, *A.Y. Jackson: A Love for the Land.*
Francine Legaré, *Samuel de Champlain: Father of New France.*
Margaret Macpherson, *Nellie McClung: Voice for the Voiceless.*
Dave Margoshes, *Tommy Douglas: Building the New Society.*
Marguerite Paulin, *René Lévesque: Charismatic Leader.*
Marguerite Paulin, *Maurice Duplessis: Powerbroker, Politician.*
Raymond Plante, *Jacques Plante: Behind the Mask.*
T.F. Rigelhof, *George Grant: Redefining Canada.*
Arthur Slade, *John Diefenbaker: An Appointment with Destiny.*
Roderick Stewart, *Wilfrid Laurier: A Pledge for Canada.*
John Wilson, *John Franklin: Traveller on Undiscovered Seas.*
John Wilson, *Norman Bethune: A Life of Passionate Conviction.*
Rachel Wyatt, *Agnes Macphail: Champion of the Underdog.*

David Thompson

Library and Archives Canada Cataloguing in Publication
Shardlow, Tom, 1958-

David Thompson : a trail by stars

(The Quest library ; 29)
Includes bibliographical references and index.

ISBN-13: 978-1-894852-18-0
ISBN-10: 1-894852-18-4

1. Thompson, David, 1770-1857. 2. Explorers - Northwest, Canadian - Biography. 3. Fur traders - Northwest, Canadian - Biography. 4. Cartographers - Northwest, Canadian - Biography. I. Title. II. Series: Quest library ; 29.

FC3212.1.T46S52 2006 971.03092 C2006-941801-2
C813'.3 C2006-940992-7
PS9426.O63Z62 2006

Legal Deposit: Fourth quarter 2006
Library and Archives Canada
Bibliothèque et Archives nationales du Québec

XYZ Publishing acknowledges the support of The Quest Library project by the Book Publishing Industry Development Program (BPIDP) of the Department of Canadian Heritage. The opinions expressed do not necessarily reflect the views of the Government of Canada.

The publishers further acknowledge the financial support our publishing program receives from The Canada Council for the Arts, the ministère de la Culture et des Communications du Québec, and the Société de développement des entreprises culturelles.

Chronology: Clarence Karr
Index: Darcy Dunton
Layout: Édiscript enr.
Cover design: Zirval Design
Cover illustration: Magali Lefrançois

XYZ Publishing
1781 Saint Hubert Street
Montreal, Quebec H2L 3Z1
Tel: (514) 525-2170
Fax: (514) 525-7537
E-mail: info@xyzedit.qc.ca
Web site: www.xyzedit.qc.ca

Distributed by: University of Toronto Press Distribution
5201 Dufferin Street
Toronto, ON, M3H 5T8
Tel: 416-667-7791; Toll-free: 800-565-9523
Fax: 416-667-7832; Toll-free: 800-221-9985
E-mail: utpbooks@utpress.utoronto.ca
Web site: utpress.utoronto.ca

International Rights: Contact André Vanasse, tel. (514) 525-2170 # 25
E-mail: andre.vanasse@xyzedit.qc.ca

TOM SHARDLOW

David THOMPSON

THE QUEST LIBRARY

A TRAIL BY STARS

For my wife, Sue

Map by Andreas (Andy) Korsos, professional cartographer

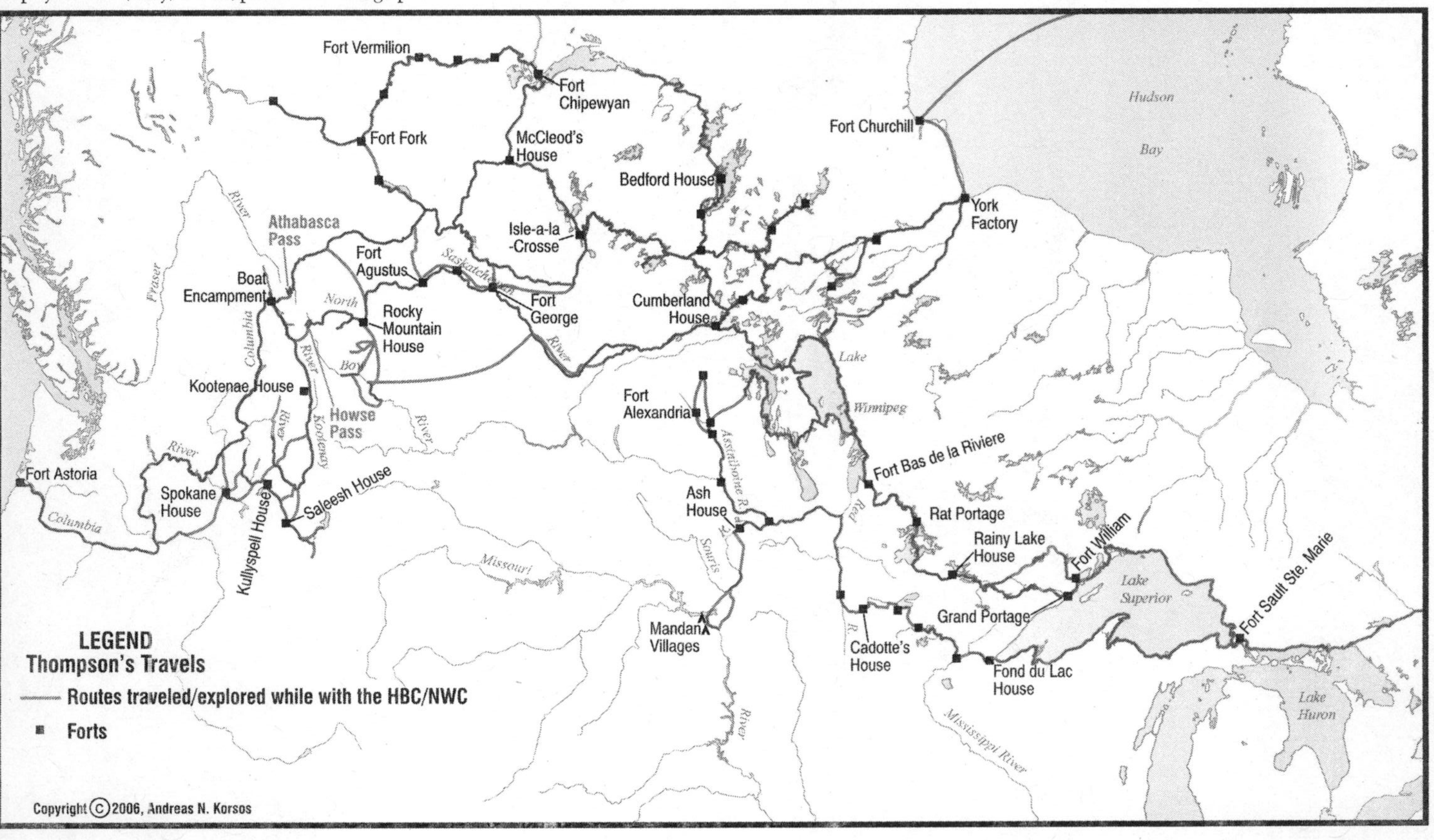

David Thompson's Travels

Contents

Library and Archives Canada/C-008711.

David Thompson exemplified the motto "Perseverance" written across the top of the North West Company's coat of arms.

Prologue

Montreal

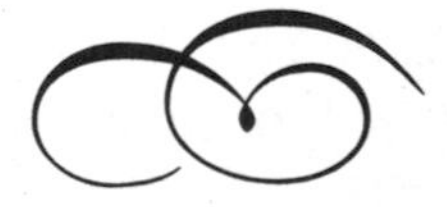

The old man takes only small tentative steps toward the general store. He is trying to keep his balance on the hard-trodden snow that covers the street. Hunger more than old age has left him weak, and he is faint though he has gone only a short distance. He hunches his shoulders against the cold and wraps his moth-eaten greatcoat closer round him. He stops to watch as a horse and sleigh pull by and uses their passing as an excuse to rest. But the jingle of bells and the cheerful tittering of sled runners in the snow seem only to mock him.

Montreal is in the lock of a winter more bitter than most. The river is frozen and there will be no new supplies until shipping opens after the thaw in April.

The city is overcrowded this year. Paupers, feet and legs bound with rags, shuffle half-heartedly along narrow roads and look for warmth. Unlike him, these are recent immigrants. Thousands arrived from Britain in summer before the ice-up. Many are fleeing the poverty and suffering of Ireland. They are poor and ill prepared for the Canadian winter and have taken up shelter where they can. Some are packed into dingy lodging houses, but most can afford only crude accommodation found in derelict hulks frozen in the river, or in empty warehouses. These are the lucky ones. Typhus and cholera have claimed the weak and malnourished. Some died while still on the defiled ships that brought them. More, having set foot in the new land, will die on the stone streets waiting for help that never comes. The old man is fortunate to have a small apartment above a shop in one of the dark limestone buildings.

When he finally arrives at the general store, he straightens himself, then reaches beneath his frost-crusted coat. He wants to be sure his bundle is still securely tucked under his arm. Behind the counter, the shopkeeper sits reading his ledger. He pretends not to notice the old man's entry and hopes the pauper will not ask for more credit. Glowing coals in a cast-iron stove warm the air and the room is laden with the smell of tobacco leaves, lamp oil, canvas, and tar. Quarters of smoked meat are suspended on hooks over rows of barrels and crates of dried and salted food. Behind the barrels are shelves stacked to the ceiling with blankets and clothing. There are no customers because few can afford the elevated prices the shortages have produced. The tattered figure approaches the counter.

"Mr. Thompson, how may I be of service?" says the shopkeeper, still looking at his ledger.

The old man does not reply but places the cloth-wrapped bundle on the counter and fumbles with its knotted ends. He exposes a brass apparatus inscribed "David Thompson, astronomer & surveyor." It is the last of his surveying instruments. He has already been forced to sell his compass, sextant, and other equipment.

"How much for this on pledge?" the old man asks.

"Perhaps if you could tell me exactly what *this* is," says the shopkeeper, "I may be able to help you."

"It's a theodolite, a very valuable survey instrument."

"Well, Mr. Thompson, I don't see there would be a real need for survey instruments here. Everyone in Lower Canada knows where Montreal is," says the shopkeeper, wishing others were in the store to appreciate the cleverness of his reply.

The old man stiffens. When he'd arrived in Rupert's Land in 1784, little other than the location of Montreal had been known about the country's geography. Using his survey instruments, he had plotted and mapped nearly four million square kilometres of North America, west of what the shopkeeper was now calling Lower Canada. He had explored the headwaters of the Mississippi. He had been first to follow and chart the Columbia River from its source to the Pacific Ocean. His sextant and compass had helped him discover and survey the Athabasca Pass through the Rocky Mountains and had been with him when he forged wilderness trails and established trading posts for the

North West Company across the continent. But he doesn't protest. The shopkeeper is right, after all. His survey tools have little value here.

"I want to settle my account and use the remainder for canned goods and cured pork."

"I'm not sure there will be a remainder. This theodolite, as you call it, not being new, I can only allow you a few shillings above what you already owe," says the shopkeeper, knowing the new railroad in the West will buy survey tools at a very good markup.

Thompson returns to the cold, unwelcoming streets with enough food stuffed into his pockets to last a few days. He regrets how the city has changed since he first travelled here in 1811. Then, Montreal was still a fur-trading centre as it had been for two hundred years. The tall ships of Europe off-loaded axes, pots, and muskets and exchanged them for pelts of beaver, marten, and fox, brought by canoes from the West. The North West Company was the city's chief trading establishment and, at the peak of its power, its influence had stretched from Montreal across the continent and reached from desolate Arctic bays to the crowded ports of China.

Thompson remembers joining the North West Company in 1797. From the beginning, it had been a loose partnership of rugged French-Canadian voyageurs and fiercely determined Scottish and English fur traders. Together, these men were a brotherhood of explorers unlike any ever known. They searched the wilderness for profitable furs and established trade routes through the vast western plains, and in so doing laid the foundation for a future nation.

There was Alexander Mackenzie, who in 1789 was first to plot the magnificent Mackenzie River from its source to the Arctic Ocean 2400 sinuous kilometres later and who, in 1793, was first to travel overland to the Pacific. There was Simon Fraser, who explored the upper reaches of the Peace River, established the first trading post in British Columbia, and who was first to map the course of the mighty Fraser River to the ocean, in 1808. These Northwesters were legends, and he, David Thompson, had taken his place among them as an equal.

In the old days, they didn't mind being derided as "the peddlers from Montreal" by their rivals in the Hudson's Bay Company (HBC). The Northwesters, led by Simon McTavish, were wealthy and spent lavishly, monopolizing the local economy and dominating the city's social life. "Fortitude in Danger" they had hollered in a toast to the company at Dillon's Tavern, their favourite meeting place. They were the lords of the wilderness and the rulers of Montreal's fur trade.

The Northwesters' victory had been hard won. Each side robbed fur shipments and burned stockades. Blood was spilled when HBC men met Northwesters along the fur trails. Canoemen were shot while paddling, and small settlements at the farthest outposts were left to starve when their caches of winter food were stolen. In 1817, the murderous competition culminated in small-scale warfare along the Red River. There, Métis of the North West Company and Lord Selkirk's mercenaries and settlers, backed by the HBC, killed each other in pitched battles with musket and ball.

Now, almost four decades later, the battles are over and Dillon's Tavern is no different than any of the city's two dozen other taverns. The fur trade has moved north, back to Hudson Bay. Montreal has evolved from its roughness into the commercial heart of British North America. In the better parts of town it is crowded, and busy shops serve well-dressed ladies. There is gas lighting in the theatres and banks. Liveried servants and French waiters give the hotels the trappings of Europe's finest. Ships from America and across the Atlantic continue to ply their way sixteen hundred kilometres up the St. Lawrence to reach the city. But now they bring fine china and expensive furniture and, for the return to Europe, fill their holds with lumber, nails, and wheat instead of furs.

The old man does not share in the new economy. His time has passed. His friends and supporters in the North West Company have long since died or moved on to other ventures after selling out to the Hudson's Bay Company. He is not, and cannot be, part of the new prosperity. His heart is still with the vast and unexplored wilderness and with the quest to discover new places and new sources for the precious pelts of beaver and marten. He longs to exchange the desolation of the city for the hardships of the trail. He prefers the companionship of French-speaking canoemen and quiet Chipewyan guides sitting around an open fire under a bright cordilleran moon to the congested streets and the company of shopkeepers and merchants. He gladly would exchange blackflies and mosquitoes, paddle-weary arms, bitter cold, and bad food for the loneliness of his Montreal lodgings, but his old age and near blindness keep him prisoner in the city.

Few in Montreal's lower streets know who he is. He makes little effort to be recognized. Some know him uptown in the financial district, but he is unwelcome there. He had forsaken the Hudson's Bay Company to join forces with the rival North West Company. Now that the Hudson's Bay Company has taken over, he is branded a traitor. Rumours persist, probably encouraged by the Hudson's Bay Company, that he is untrustworthy and has betrayed the colonial powers by stalling exploration of the Columbia River. This, it is rumoured, helped the Americans successfully expropriate the rich Oregon Territory from British hands.

Near the docks he finds the stairway leading to his modest lodging. He ascends, placing each frail step deliberately before the next. His Métis wife, Charlotte, is waiting in an unheated room. He sits beside her on the narrow bed and between them they examine the meagre supplies he has brought from the store. This is not the first time they have been hungry. Over the years, Charlotte has shared the adversities of the trail with him, and she has known the hungry nights when the pemmican was gone and the game was scarce. Those hardships, even though time has softened the memory of them, are still more bearable than this. She knows how deeply it hurt him to pawn his precious instruments. On the trail, only he had been allowed to handle them, and she had watched him hold them as if they were more valuable than all the pelts they carried. She understands that his instruments made him more than just a fur trader, they made him a map-maker like Captain Cook. Using them, he was able to chart, with paper and ink, the rivers and lakes to the great ocean in the West. Paper maps are of little comfort to her now, and she is afraid this final suffering will break him and he will die soon. Then she too will die because she is old and cannot return to her people on the plains.

Still in his coat, the old man shuffles to the table underneath the room's only window. A muted light penetrates the frosted pane and allows him to read his pen-scrawled journal. He rubs his hands together, then, taking up a quill, he makes a sharp jab to break the thin film of frozen ink in the well. He transcribes another coordinate from his journal into his latest map. Almost blind, he can barely write. The gentlemen who had agreed to pay for the publication of earlier maps have found more profitable enterprises, but Thompson is still hopeful. Perhaps if he can sell his maps and maybe even his journals, then Charlotte and he will be able to make do until spring.

David Thompson would survive this winter, and the next, and several after that, but early in 1857 he died penniless and in obscurity. He was buried in Mount Royal Cemetery, Montreal. Charlotte died three months later, and her body was placed beside her husband's. Their graves were unmarked.

1

London

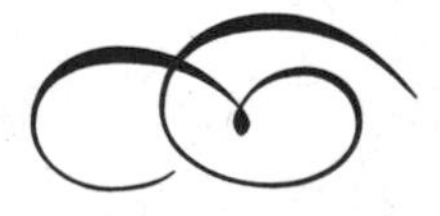

At 7:30 a.m. the portable gallows was wheeled into position. Hangings were always conducted by 8:00 o'clock in the morning in London and the execution, like many in the year of 1783, was to be carried out near the place of the crime. Even though the event was still a half hour away, the market square was already in a festive mood. Jugglers and street musicians were plying their trade. Elegant coaches arrived and took up the favoured position nearest the gallows. Rooms overlooking the square were temporarily rented to those who could afford them, and ill-mannered onlookers leaned from dozens of windows. Street vendors sold buns and drinks. To one side a group of young rakes entertained the crowd with a mock hanging as they

Spencer Art Museum, University of Kansas/*Gin Lane* by William Hogarth.

He never forgot. At an early age Thompson was exposed to London's appalling poverty, made worse by the effects of cheap gin.

mimicked strangulation. Just before 8:00 o'clock red-coated soldiers marched in and cleared the way for a plain wooden cart with an open top and side-rail pickets. In the cart sat a woman with her hands tied to the pickets, and beside her was her coffin. The procession rumbled up to the scaffold. The soldiers untied the condemned and led her to a position under the noose.

This was the execution of Judith Dufour, who they said murdered her two-year-old daughter. It could just as easily have been the hanging of some hapless chimney sweep, who might have done nothing more than steal a sausage from the butcher's shop. The noose gave little regard to gender or age and accepted anyone who was led up to the elevated platform. On average, in the 1780s, there were two public executions a week within the confines of the city. This was not surprising since there were 350 offences for which one could be hanged in the late 1700s.

On the platform stood the hangman, a clergyman, and a sheriff. The sheriff was presiding as the speaker from the courts. The crowd became still as he, reading from his document, shouted to the assembled. "It is the order of His Majesty's court that Judith Dufour, found guilty of the crime of murder, be hanged by the neck until dead and thereafter her body is to be buried within the precincts of Newgate Gaol and may the Lord have mercy on her soul." Judith, it was said, had left her daughter at the poorhouse, where the infant received new clothes and food. The mother later had returned for her child and killed her after stripping her of her new clothes. She then had sold the tiny garments for one shilling and fourpence worth of gin.

"'Ang the bitch!" someone shouted, inciting jeers and demands for justice from the clamouring crowd. Judith, with her wrists bound, was still able to unpin a small crumpled hat from her hair and carefully hold it in both hands. A white hood and noose was fitted over her head as the clergyman murmured his brief recitation. That done, he stepped back. On the sheriff's signal, the trap door sprang open. There was a brief silence, then roars of approval as her body twitched violently at the rope's end. Beneath her, a minor scuffle ensued. Souvenir seekers shoved and jostled, hoping to snatch her fluttering hat.

Not far from the gallows and Newgate Gaol, where the condemned prisoners were kept, David Thompson was busy cleaning windows and washing walls. Here, next to Westminster Abbey, was the Grey Coat Charity School for orphaned boys. David, a pupil at the school, was doing his morning chores. Grey Coat was a rare haven from the often brutal life of London's poor. Most orphans were street urchins, stealing and scavenging to eat and hiding to escape the law. Without family support they were guttersnipes and fell easy prey to London's unsavoury underworld. David, poor and without a father, was among the usual candidates for the gallows, but at Grey Coat, charity had intervened.

The boy washed smuts from the outside windows and front door of his school. Armed with a rag and a pail of cold water, he carefully scrubbed the door's wooden panels. He even worked his wet cloth into the corners of the iron hinges in a futile effort to remove the stains. The insidious black fallout from the city's innumerable coal fires coated everything. It darkened the stone fences, iron gates, and headstones of church graveyards. It coated the walls of Westminster Abbey and settled on the ramshackle hovels of Whitechapel. The soot leached from wet cobble of the market square and drained into black pools in dark alleyways. It shrouded the palace rooftop and stained the backs of white horses pulling fine carriages. It settled on docks and painted warships tied in the river. Everything was blackened. On bad days the city's airborne fume even crept into the school and into the fabric of his clothing.

Worse yet, it blotted the pages of his schoolbooks. After wiping each windowpane, he wrung his rag out onto the cobble. He was careful to keep the splash, now mixed with the street's horse dung, from soiling his newly washed uniform. But this was impossible. The small thirteen-year-old boy was jostled and nudged in the crowded street by a troupe of jugglers rushing from the market square. He spilled dirty wash water onto his white leggings, yet still he hurried to finish the windows. Now he needed enough time to wash out his clothes before the school assembled for the evening meal.

Cleanliness was essential at Grey Coat. Keeping clean prevented the cane, and anything that helped avoid a painful beating from that stiff rod was worth the attempt. The "rod of chastisement," as it was sometimes called, was not really a rod or a cane, the orphans reasoned, trying to comfort themselves, but only a dried willow branch. It measured a metre long and was just thicker than the big finger on the headmaster's podgy hands. The cane was worn smooth from its use on the backs and hands of schoolboys who were found unclean or who broke a rule. And yet, this school may have been only slightly less forgiving than other Charity Schools in the year 1783, and David knew he was fortunate to be here. These few boys, including David, must have felt transported, as if by divine intervention, into another world.

Within Grey Coat's secure walls, the boys were able to share in the best their era had to offer. Under the tutelage of the school's knowledgeable clergy, they found themselves to be living in the age of the Enlightenment. They learned of famous Englishmen like Edward Jenner, who had originated a technique he called vaccination, which prevented the dreaded disease smallpox. They heard about the astronomer Edmond Haley, who predicted the coming of comets and forecasted heavenly events that, until then, were only known to the Almighty. They knew James Watt was developing an engine powered by steam. Thomas Adams, a favourite teacher, took them to street demonstrations like the one on the new phenomenon called electricity. The boys saw the first simple generators produce loud cracklings and sparks that ignited trays of heated spirits to the cheers and amazement of onlookers.

Grey Coat's training emphasized mathematics, astronomy, and maritime studies used for navigation. The school was preparing its students for service in His Majesty's Navy. The boys were taught about the Royal Society and they learned of John Harrison, whose chronometer was revolutionizing navigation. The students were enthralled by the accounts of Captain Cook who, with sextant and compass, had recently charted unknown Pacific waters. They fantasized that one day, they too might be part of some new exploration. But their reality was, at best, a future in the navy. The navy was in constant need of replacements for those lost in battle, and it was difficult to recruit sailors with enough education to help navigate a fighting ship. David's chances of becoming an explorer were as remote to him as the throne of England.

Before supper there was an assembly for inspection, and the headmaster examined each boy to look for unwashed hands or soiled shirts. Either infraction could mean no supper and often a caning for good measure. Discipline at the school was harsh, but applied according to the customs of the day. After supper, the orphans sat studying at wooden tables. This was where they spent most of their time when they weren't cleaning windows, sweeping halls, or scrubbing potatoes. They had little time to do anything else, but sometimes, secretly, using a ruler as a cutlass and a candle as a pistol, they played at quelling mutinies and skewering Spaniards. Some were destined to become midshipmen and would have a chance, although a remote one, of becoming junior ship's officers like some other Grey Coat boys before them. They were told the King needed trained officers to repress the revolt in the American colonies and to quiet uprisings in India. They would be ready.

For David, now in his final year at Grey Coat, these promises of the navy seemed distant. It had all changed, he remembered, on that cold morning last January. David and Sam had stood nervously in front of the headmaster's desk. Neither boy was sure why he had been summoned. David half expected a caning, and his mind raced over recent events as he tried to think of what infraction he could have committed. They were seniors, and he knew they would be sent off to sea by summertime. But that was still months away. Just the same, he quietly hoped this was their call to join the navy as midshipmen or, even better, as expedition crew members.

"Master Thompson. Master McPherson," the headmaster barked from his high-backed chair. "The Navy Board has informed me that with peace again this year, they will not be in need of Grey Coat boys at this time. However, the school has made a generous payment of five pounds sterling to the Hudson's Bay Company for each of you. I have accordingly arranged that you both shall be taken into that company's care and service. You will be apprentice clerks to the fur trade and posted in the New World for seven years. I trust you will find Rupert's Land agreeable and will represent your school well."

The headmaster's words could not have been more disappointing. David was heartbroken, but he knew there was no use in complaining. He remained silent.

"There is plenty of time to prepare before you are to be taken to the Hudson's Bay ship, *Prince Rupert*. Your vessel doesn't depart till June, which is six months from now."

The headmaster rose from his chair, reached forward, and briefly shook both boys' hands. "Be sure you are ready and God be with you."

Six months was a long time to brood over their disappointment. For Sam, it was too long. He had slipped off quietly in the night, choosing the perilous life of a London street orphan over the uncertain fate waiting for him on the frozen shores of Rupert's Land. David knew the cruel streets of London as well as any of the boys did. He hoped Sam would change his mind and somehow find his way to the *Prince Rupert* before she sailed. David put these memories and these hopes from his mind and returned to his books.

ꕤ

There were only eighteen to twenty days out of a year that David and the other boys were allowed time of their own. Most often David spent his free time in the old abbey at Westminster where Grey Coat housed its students. There, in one of the many ancient stone-walled cloisters, he could curl up and read. He loved reading adventure books like *Gulliver's Travels*, *Arabian Tales*, and *Robinson Crusoe*. Books like these were scarce, and the few the school had were only there by the good will of charitable donors.

From the abbey he sometimes walked to nearby London Bridge or St. James's Park. If he felt adventurous, he could find his way across the market square in the morning when the rumble of wooden handcarts and the crow of roosters were just beginning. Vendors would be starting to set up stations with live poultry and fresh produce stands. He could bump past a cheesemonger arranging his stall and, leaving the market, slip down narrow lanes. He could wander toward the docksides where the morning fog from the river was pungent and still held the scent of the night before. It was thick with the smell of alcohol and stale tobacco that drifted from empty drinking cellars and alehouses. The stench of urine seeped out of dark alcoves and slatternly lanes. The docksides were a place of violence and misery, notorious for drunkards, thieves, and press gangs, but early morning would find most incapacitated or asleep. Yet, some would lay on doorsteps and others still swayed and faltered aimlessly in the street, victims of "Kill Grief" or "'Comfort," as the cheap gin was called.

These docksides and the gauntlet of drunkards and cutthroats were dangerous places for boys like David. Sam McPherson was probably here now, alone and hiding somewhere in a back alley. Maybe David's schoolmate had already been abducted by a press gang looking for a ship's boy. Maybe he was murdered the first night he ran away or maybe he was kidnapped and sold to a Mollie house where men dressed as women. Here brutish sailors aroused by drink, or finely dressed noblemen numbed with opium were using homeless boys to satisfy their sexual appetites. David and the other orphans at Grey Coat knew they had little protection once outside.

David stayed close to the relative safety of Westminster and away from the drunken masses in some of the city's poor districts. His mother was most likely somewhere among them. Unable to support her children after his father died, she had been forced to make some difficult choices. Although he missed her, he was thankful she had given him, at age seven, to the school, and he prayed he would never fall victim to drink.

The curse of alcohol seemed limitless to David, and in some of London's worst districts like Whitechapel or St. Giles, he would have difficulty finding anyone sober. Watered-down wine, beer, and poorly distilled gin were a daily diet and sometimes the only sustenance taken by the city's poor. In St. Giles one in every four houses was a gin shop. These were open to anyone of any age, and David could see men, women, and children drink themselves into oblivion. That alcohol had become a plague was obvious, but still its use was widely endorsed. The visiting physician would prescribe spirits for sick Grey Coat boys and, often as not, would take the cure in liberal amounts himself.

Alcohol was prescribed for all manner of ailments. It thinned the blood when it was deemed too thick. It warmed the bowels and aided in digestion. Spirituous liquor warded off the flu and cheered melancholia. Drink was also a mainstay of the British fighting forces. It fortified the disposition and helped soldiers quell their fear of battle. The navy's daily ration was eight ounces of grog, and sailors were sometimes given double rations before naval engagements. Many seamen were so dependent on their daily ration of rum and water that they re-enlisted into the dreadful hardships of the lower decks just to find a steady supply. To the rest – the poor masses of that crowded and dirty city – alcohol was an escape.

David was thankful for his school. There he could find his escape, not into alcohol but into the *Epitome of the Art of Navigation*, a large book from which he studied trigonometry and the techniques used to plot a ship's course. David worked hard at trigonometry. He knew that without it he couldn't navigate, and the ability to navigate a ship might still be his passport out of the city's poverty. Even though he was going to a Hudson's Bay Company apprenticeship, if he could navigate, he might yet make it to the navy's upper decks. His only other choice was to run for it like Sam and maybe find a kind skipper of a merchant ship to take him aboard.

2

Rupert's Land

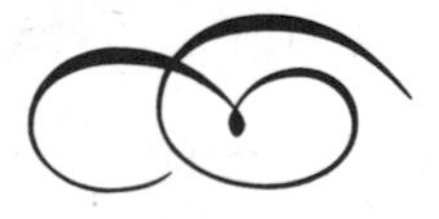

The day David had hoped would never arrive was here. He placed his grey uniform on his cot and gathered his meagre clothing and belongings. He rushed down the hall and burst into the classroom where Thomas Adams sat alone reading the Bible.

"I'm being sent to Rupert's Land today and must leave straightaway!" he gasped.

"So I've been told," replied the teacher, putting his hand gently on David's arm. "Don't be too disappointed, lad. Keep working hard. You never know what good advantage this setback may bring. Now quick to the kitchen and see what the cook has for you. Goodbye, Davie, I'll miss you."

"Goodbye, Sir." David raced to the kitchen.

Hudson's Bay Company Archives/Archives of Manitoba.

Nonsuch, the famous ship that in 1669 returned to London with a fortune in furs. The voyage spawned the Hudson's Bay Company and the creation of Rupert's Land.

Library and Archives Canada /C-082981,

A time of excitement at York Factory. Hudson's Bay Company ships offload trade goods and take on the valuable cargo of fur bound for London.

The cook slipped him a biscuit and a little salt pork. "For the road. God bless you lad."

David followed at a quick trot behind his escort. The rough-looking crewman was in a hurry to return to his ship. His instructions were to bring back two apprentice boys, but one would do. He didn't want to keep his captain waiting. "Captain Tunstall he wants us aboard early he does. Cap'n don't like things left t' last minute," he warned, as David began to lag behind. The *Prince Rupert* was rafted on the outside of two coal barges on the Thames. It was a fine Hudson's Bay ship, and its bright gunnels and new rigging seemed a good omen to David.

"There should be two boys!" Captain Tunstall insisted from the quarterdeck as the crewman and David climbed the shipside ladder.

"Bloody hell!" the crewman hissed. In panic he turned to David, demanding an answer. "Where'd the little bastard go!"

"Don't know sir, he's been gone six months," David answered carefully.

"Damn, damn, damn!" the captain growled. "Back to the school and see if you can't find where the mutinous beggar went. And you, my young fellow!" he ordered, "follow the mate to your quarters below."

David followed obediently, hoping the crewman might somehow find Sam. *The headmaster would never tolerate a runaway and even less a deserting apprentice*, David reasoned. *Maybe the headmaster knew of Sam's whereabouts all along.*

Captain Tunstall ordered that David be kept in his quarters until the *Prince Rupert* was put to sea. The second mate led him to the lower deck.

"Not a proper place to sling a hammock," the seaman confessed. "No portals or vents, but 'tis dry and warm enough. Mind don't soil the sailcloth with your shoes or the master sailmaker 'll stitch ye to the top gallants," he said, latching the door as he left. David sat silently and listened to a chorus of rumbles and shouts while the last of the *Prince Rupert's* provisions were stowed away. Eventually he struggled into a hammock and half-heartedly ate some of the cook's hard biscuit. He saved the remainder in his coat pocket, but there was faint hope of sharing it with Sam. The scent of Stockholm tar and new sail-canvas filled his nostrils as he drifted into sleep inhaling the ship's stale air.

Next morning he woke to the clanging of the ship's bell and the sway of his hammock. He was still alone, and the *Prince Rupert* was rolling her way to open water. "Show a leg! Out and down!" a great and terrifying voice called. "Up all hammocks!" the first mate ordered, and David rushed to his feet.

"Seven bells.'Tis morning watch in half an hour, lad. Quick. Come to the table," summoned a gentler voice, as the ship's cook peeked around the bulkhead and handed David a bowl of thick porridge. "Take a place at that last table there," he said, motioning to a crowd of men already huddled over their gruel at the mess tables. David and the men ate in silence, and no heads lifted to acknowledge the newcomer.

"Third watch on deck!" the first mate's command boomed down the main hatch. David raced to finish his breakfast as the crew scurried to place their bowls in the cook's washtub. He joined the rush up the main hatch ladder. On deck, each seaman hustled to his station while David remained awkwardly wondering where to go.

"To the holystones!" ordered the captain. A seaman jumped smartly into action. David watched him take a bucket of sea water and a pumice stone and begin to scrub the forecastle decks along with several other men already on their knees. He followed the men's hands, scrubbing in small circles, grinding the wooden deck into milky grey puddles until the sweeping of their hands melded with the dizzying roll of the ship. Before the *Prince Rupert* entered the Channel swell, his stomach was knotting in nauseous pulses. He fought the sensation until involuntary convulsions finally sent him for the rails, where he spilled his porridge into the English Channel.

"Seasick? N'er mind," jibed a seaman while continuing to scrub. "The admiral himself spends some time bent over the rails till he gets his sea legs. Welcome aboard, your bloody Lordship," he scowled. David leaned over the railing again to the laughter of the crew and spent the remainder of the watch imitating the admiral. At nightfall, he fell exhausted into his hammock.

The next day he had thankfully given himself to the motion of the ship and was put to work doing some of the normal shipboard activities like cutting sail patches or scraping varnish from weathered yards. Toward afternoon, a small ship sailed to within hailing distance. It was a Dutch lugger, and by the shouting between ships, David understood the Dutchman was selling contraband gin. The ships steered off, putting a safe distance between them. The *Prince Rupert*'s gunner and four eager deckhands climbed into a boat that was lowered over the side. Smooth and synchronous strokes on the oars had them to the lugger in a short time. On board, the gunner was quickly handed a taste of gin from a bottle snatched by the Dutch captain from a full case. "Hurry!" said the captain in rough English. "De Revenue Cutter he's cruising near at hand! Ve must luff off. You must go!" The gunner hastily paid a guinea for the whole case and the men loaded their prize into the launch.

"Better 'n pay'n London prices," remarked the second mate on the *Prince Rupert*'s main deck as the precious cargo was hoisted aboard. The case of gin bottles clanked safely onto the grating, but the satisfaction of the gunner was soon spoiled when the old carpenter suspiciously uncorked a new bottle and took a short swig. "Sea water!" he spat. "You try another."

"All bloody sea water!" cursed the gunner as he spat out another mouthful. But it was too late. The Dutchman had gone on a fast tack more than a kilometre off *Prince Rupert*'s stern. The gunner and his men were in a fighting mood, but Captain Tunstall ordered a steady course. Although the crew would be irritable and restless without their gin ration, he wanted to put in at Stromness by June 1. The latest Company dispatch from London would be waiting. It contained his final instructions before their Atlantic crossing.

David's mood lifted as sunrise revealed the hills of Scotland lying blue on the horizon. This was his first glimpse of anywhere outside of London. The crew scrambled aloft and let out sail at the mate's sharp command, and the *Prince Rupert* rolled into the steep choppy seas off the Islands. A fresh wind kept them tacking as they laboured to windward until nightfall.

At 9:00 p.m. the dark silence was broken by the clattering chain as *Prince Rupert*'s anchor plunged toward the muddy bottom of Stromness harbour. David found the stillness of the harbour a relief after the constant sway and roll of the last six days and nights under sail from London. He climbed into his hammock wondering what the light of next day would reveal about this quiet place. He was wakened in the morning by smoke. By its smell, it was not from the galley fires, but an acrid sweet smoke from some different fire. David went on deck to investigate and was struck by a barren landscape.

"No trees," he said aloud to no one in particular.

"That's so people here won't spoil their clothes trying to climb them," a sailor answered.

On the treeless shore, David saw the smoke's source. Five stone kilns were bellowing black clouds from smouldering seaweed as the wet green harvest was rendered by fire into fertilizer. He watched men and women struggling back and forth on the distant foreshore, carrying baskets of dripping seaweed to the kilns. Somehow harvesting seaweed made sense here, for nothing else but short green stubs of grass seemed to grow on the rocky grey landscape. Even the tiny cottages dotting the shoreline of small islands in the bay were made of rock and had sod covering their roofs.

Other ships were anchored in the bay. All were Hudson's Bay vessels waiting at this remote station tucked into the Orkney Islands on the west coast of Scotland. They too waited for the last dispatches from

London. The village of Stromness was a customary stopover for Hudson's Bay ships bound for the New World. Over the next few days they would take on water, buy fresh-caught herring, and stow away their final provisions. At the head of the bay, just beyond two small islands, was a friendly cluster of small brick houses with slate roofs. Smoke from peat fires rose lazily from their chimneys and drifted down over the stone seawall at the harbour's edge. Behind the village, the hills lay low and misty.

At a warehouse, *Prince Rupert* safely bought genuine contraband from the steady supply smuggled into Stromness harbour from Holland. The crew loaded several kegs of "Comfort" for the cold Atlantic nights. Late that afternoon, Captain Tunstall had the other captains and some gentlemen from the island brought aboard the *Prince Rupert* to dine. As they prepared to retire to the captain's table the wind shifted and pungent smudge from all five kilns drifted across the bay, blanketing the ship's deck. Day was turned suddenly to night as the suffocating fumes enveloped *Prince Rupert*'s dinner party. Captain Tunstall ordered the boatswain ashore to demand the kilns be extinguished at once. The boatswain, however, was met with an adamant refusal by the islanders.

"Then we shall turn the ship's guns on you and have our cannonballs put an end to your kilns," threatened the boatswain.

"You may as well take our lives as our means," they said stubbornly, "We will not put them out."

The boatswain had dealt with Orkadians before. They were as tough and hard as the islands on which they scratched out a living. He would have to try something else.

"How much do you make a day from these kilns?" he asked.

"Ten pence."

The boatswain reached for his money pouch.

"Each," added a soot-faced kilnsman.

The boatswain handed each man a shilling and the kilns were out before his launch had rowed back to the ship.

They set sail for Rupert's Land on July 3. Over the next few weeks conditions grew steadily worse. North Atlantic squalls battered the ship. Treacherous icebergs, looming and ominous, threatened to rip through her belly timbers. The food became maggot-ridden and foul. But maybe worse than maggots, for David, was the ill-tempered crew. Mistrustful of anyone who didn't drink, they excluded him like the runt in a wolf pack, and he was as alone as he had ever been. Even the captain, whose attention David must have longed for, seemed not to notice him at all.

He consoled himself by observing the stars and the wind. With daily readings sneaked off the binnacle, he tried to guess the ship's course. David desperately wanted to help navigate the ship. He had learned the skills for it. He was sure he could calculate *Prince Rupert*'s position to within a minute of latitude, given a chart. But that was not to be.

August 30 finally brought a calm evening after nearly two months of heavy seas. David, alone on deck, searched skyward for Polaris, the North Star. *Prince Rupert* was nearing land, some crewmen had said. David reckoned, as best he could with no chart or sextant, that they were well into Hudson Bay. The day before, he had overheard the master's mate call out their position as 59 degrees and 03 minutes north latitude. Being able to find his way was everything to him. It meant he could not be lost, would never be left adrift not knowing where home was. His schooling had taught him that no matter where he might be on all the world's seas, he could know his position by the stars and by degrees and minutes. But he could only guess where he was now, and he would likely never navigate a ship. He was destined to be a clerk, to waste away counting blankets, buttons, and company tokens in some forgotten outpost. The prospect of his bleak future, the strain of the voyage, and the memories of London, were being held off by great effort and were about to crush him.

"Fine evening, Master Thompson," Captain Tunstall remarked as he tugged indifferently on the mizzen halyard, his back purposely turned to David.

"Yes sir," recovered David.

"Watching the stars again, I see."

"Aye sir."

"How far to Churchill by your reckoning?" asked the captain.

"Two days sir, give or take."

"So it is, so it is," nodded the captain. "Two days with this southeast wind in our favour." The captain paused, giving David ample time to speak, then turned his gaze seaward. "I've spent many years at sea and have seen many a man and boy pass over my decks," said the captain, as though thoughtfully addressing the waves. "Most were good men, all and all. Oh, they had their frailties, mind you, some for gin, some for worse, others just lazy. And some, some was lower in nature than hagfish – mean and hateful. And I can say now that the measure of these men was not known 'til seen in adversity. The sea throws up her hardships sure enough. And some can weather a storm, float'n easy like petrels fly'n in a gale. Others, loud and brassy in the calm, why they can crawl into themselves and shrink from the wind and fury like a snail hiding in its shell. Aye, and there's no telling. I've seen some starts like snails and become stout hands and others, veterans of the bloodiest battles on land or sea, take to trembling uncontrolled at just the sound of a hatch slammed shut."

"It's not bloodlines. Nobleman or foundling makes no difference. God knows our very own King is bound for Bedlam. They say Mad George is at present strapped to a gurney and wailing at the moon. If he were on this ship I'd be putting him in leg irons for his own safekeeping. Half the Royal Family is mad. Sheltered too long, I think, from real hardship and lost the ability to cope. But it's not hardships either that makes the man, David. Why if that were so, you are all the man you ever needs be now. What worse can befall a boy than to lose all his kin, hav'n everything safe and familiar stolen from him when he needs it most? Now you stand here hale as any, yet that other lad, same age, same Grey Coat boy, same orphan, he ran."

"No, I'm think'n it's like each man's heart is a bucket with a slow leak. The heart can hold only so much trouble and misery. Given time, disappointments will gradually drain away. But too much and it fills a heart to flood'n over and not the bravest man can stand it. He brims over with anger, or hate, or shuts himself off, or shrinks away, each in his own way. But by God, Master Thompson, I can see you've already grown a barrel of a heart, lad. Stay open to it, keep it big, and you'll make this New World yours, all right. Aye, and I wouldn't be surprised if you'll be the best damned apprentice the company's ever put ashore."

On September 2, 1784, *Prince Rupert* slipped her way into the wide mouth of the Churchill River. They sailed past the ruins of Prince of Wales Fort where charred granite walls and rusting guns stood as mute guards on the low-laying north bank. The fort, destroyed in the war with France, was never rebuilt. Captain Tunstall steered his ship on the rising tide, eight kilometres farther upriver to the new post at Churchill Factory. His destination reached, he ordered the factory's provisions offloaded. This done, he commanded, "Make ready for loading," and the crew began to take on the forty-kilogram bales of pressed beaver, muskrat, and marten pelts, which were stacked on the dock ready for London's fur market.

The captain gave David a friendly nod as the apprentice disembarked to the longboat. Ashore, David followed the well-worn footpath that led to the factory. Part way up, he turned and stared back at the ship. He put his hands into his coat pocket and discovered the hard biscuits and the rock-hard piece of salt pork given to him by Grey Coat's cook five months earlier. He tossed them into the marsh grass by the trail. Then, as hungry gulls descended noisily on the offering, he walked up the path to the factory.

No friend of Thompson. Samuel Hearne, the embittered Hudson's Bay Company chief factor at Churchill.

Hudson's Bay Company Archives/ Archives of Manitoba.

Library and Archives Canada /C-073431/C.W. Jefferys.

Guns for fur. Firearms and alcohol would forever change the way of life of the Aboriginal Peoples.

3

Churchill Factory

Within ten days the *Prince Rupert* was loaded and ready for London. She could not delay. A cold arctic front was blowing its icy breath over the river. Northward, sea ice was already starting to block navigation out of Hudson Bay. From the rocky headland, David watched *Prince Rupert*'s longboat tow her into the out-flowing current where the ship drifted quietly downstream to the sea. *My ship could be off Dover's white cliffs*, he thought, *within the number of weeks counted on one hand, but I've no chance of leaving this place for seven years at least, maybe never.* When the *Prince Rupert* was finally out of sight, he felt he had lost all hope of contact with England, with his school friends, with his childhood.

He wandered back to the unfurnished bunkhouse and pulled a rough wool shirt from under the wooden planking of his new bed. The morning sun seemed to give him little warmth, and he buttoned on another layer under his coat and joined his new workmates, two company clerks, who were collecting firewood. The three of them had been scouring the river for days, hauling driftwood up the bank, chopping it into stove-size pieces, and stacking it inside the factory's stockade, which was nearly four metres high. Wood was scarce. Over one hundred years of occupation at the fort meant even the smallest of the stunted trees had been stripped away. The already naked landscape had become lunar-like, emptied of all wood and scrub. The river floated a fresh supply of driftwood downstream each week, but that was still not enough, and the work parties trudged many kilometres each day in search of more firewood.

For David, the weary chore of wood collecting was interrupted only by joyless clerical duties. With ink and quill he made entries into the Hudson's Bay Company ledger: 7 forks, 12 balls of twine, 118 small bags of gunpowder, 20 eight-lb. bags of bird-shot, 4 one-lb. packets of salt, and on it went until every item in the multitude of items unloaded from *Prince Rupert* were packaged and stored into the company's inventory.

Pelts of muskrat, fox, and beaver brought in by Chipewyan trappers would be exchanged according to strict standards established in London. One blanket had the trade value of forty muskrat, or two black fox, or seven prime beaver. The post's chief trader, called the factor, added his own markup, which in some cases doubled the number of furs required by London for a trade item. This was called the Factor's Standard. Each night the inventory from the previous day's trade was tallied and recorded in the company's books. The factor's take was recorded in a separate book.

"Still at it this late at night?" a voice called through the storeroom door as David sat reading a borrowed book. Through the dark room he watched a slim-built man approach and reach his open hand into the dim light surrounding David's candle.

"Hello. I'm Hodges, company surgeon, pleased to make your acquaintance."

"Thompson," David replied, shaking the surgeon's hand. "Pleased to meet you, sir."

"Mr. Prince and I are off hunting grouse tomorrow. We could use a third if you'd care to join us?"

"I'd like that, sir, very much so Mr Hodges, but I have firewood to collect and the factor gave me his maps and journal to edit."

"Don't worry, I've already cleared it with the factor. A lad can't spend all his time bent over books."

"But I don't know how to shoot," David admitted, then wished he hadn't. The offer to get away from the boredom of the post was too good to miss.

"Well, then, it will be my pleasure to teach you. You won't be of much use in this country if you can't shoot. Meet us at the stockade gate at sunup," said Hodges cheerfully as he left.

The day broke grey and cold. The first snow of the year drifted lazily on the wind and covered the rock-strewn peninsula with a dusting of drifting white powder. David followed Hodges and Mr. Prince from the factory gate to a dinghy beached on the riverbank. They rowed past the small HBC sloop *Charlotte*, anchored in a quiet back eddy out of the main current. Mr. Prince, the captain of the HBC boat, had brought her in a few days earlier. The *Charlotte* and her three-man crew would be wintering at Churchill, as they usually did upon completing their trading duties with the Inuit people in the North. The hunting party nudged the dinghy's keel into the sand of the far bank and hauled the little boat high up to the grass. Here, Hodges handed David a long, slender musket and explained, "This is a flintlock, Master Thompson. You pour exactly this much gunpowder down the barrel," he said, tipping a powder horn into the gun's muzzle, "and then ram a cloth patch, like this one, down to hold the powder in place."

Hodges helped David pull the long ramrod from under the gun barrel and with it push the patch deep into the barrel. "Now," said Hodges, as he put a cluster of small lead pellets in David's hand, "pour a little bird-shot down the barrel and ram down another patch to hold the shot in place."

David followed the instructions carefully. "Good," said Hodges. "Now you need to put a small powder charge in the pan on the side of the barrel and pull the hammer back like this. When you pull the trigger, the hammer holding a piece of flint will strike the frizzen. This makes a spark, which will ignite the pan. The pan's powder will burn through the flash-hole and ignite the main charge in the barrel, which will explode and force the lead pellets out the front end of the barrel."

David's eyes were beginning to glaze over when Hodges added, "But don't pull the trigger just yet unless you want to shoot Mr. Prince in the foot, which you happen to be aiming at."

David sheepishly pointed the musket away from the captain's muddy boot. He raised the stock to his shoulder, aligned the sights with the distant horizon, and pulled the trigger. A flash of fire and smoke arose from the pan, and an instant later the gun's recoil slammed into his shoulder when the charge roared out the barrel, leaving a second cloud of black smoke.

"Good!" shouted Hodges, as David staggered to regain his balance. "Now you know how to shoot. Just repeat the procedure, but next time, hold the musket steady while pointing in the general direction of a grouse. We'll have lunch in no time."

Fortunately for David there were plenty of grouse to practise on. By day's end he had managed to bag two ptarmigan, or grouse as they called them. That was his contribution to the day's total take of thirty-five birds.

Samuel Hearne, the chief factor at Churchill, was waiting on the river's bank as the hunters returned. Hearne was just entering middle age, but was more than ready for retirement. His long hair had begun to grey, his shoulders sloped forward, and he stood as if being pulled toward the ground by a force of gravity stronger than that felt by other men.

"Gentlemen," greeted Hearne. "How went your day?"

"Thirty-five birds in all," answered Hodges.

The factor nodded. "We'll need all the game we can store away. Our beef supply from London is not generous this year. How did the apprentice do?"

"Damned fine," beamed Hodges, "He's a natural for shooting."

"Then have him hunt with you, and if he's good enough with that fowling piece by the time the geese are migrating south, take him to the marshes and stock up on as many geese as you can shoot," he ordered.

David hardly had time to enjoy the good news when he noticed Samuel Hearne staring at him in the way one might look at a badly behaved dog. David suddenly felt very uncomfortable.

"See me in my quarters tonight at eight, Master Thompson," said Hearne sharply.

Some days before, the factor had given David a scrawled journal and some rough maps to edit. The untidy manuscript was Hearne's first attempt at describing his extensive journeys to the Arctic. He wanted his journal to be readable and his maps made ready for publication. But David's editing made it clear to Hearne, and to others at the trading post, that the boy's education, his grasp of map-making, and his ability to write was superior to that of the chief factor.

David hunted the next day with Mr. Prince and Hodges, but he was sullen.

"How was your meeting with the factor last night?" asked Mr. Prince as they sat on the crumbling stone ruins of the old fort, resting from the morning's hunt.

"I don't want to talk about it," said David, who knew it was not his place to find fault with the factor.

He never did disclose what happened between them. Did Hearne berate the new apprentice for daring to criticize the geographic accuracy of the factor's journals? Had Hearne, known to indulge heavily in liquor and keep a number Chipewyan girls in his residence, shocked young David's sensibilities? Did the factor, who dismissed Christian beliefs as dogma, threaten the boy's deeply held religious values? The answers to these questions will never be known, but from that time on David Thompson would harbour a lifelong dislike of Samuel Hearne.

"Old Hearne, he's a changed man these days and I fear not for the better," said Mr. Prince, surmising that there had been trouble with the factor the night before. "His defeat two years ago, at these ruins we now sit upon, seemed to unravel him. He's never been the same since. He had settled into the comforts of the fort after years of hard travel. He made three remarkable expeditions north, as you may know by now from reading his journals. His last foray was all the way to the Arctic Ocean. It was 3500 miles round trip. He covered unexplored territory nearly the size of Russia to look for new sources of fur for the company. He proved there was no Northwest Passage to the Pacific out of Hudson Bay, but the authorities in London wouldn't believe him and further insulted him with a measly two-hundred-pound bonus for his troubles.

"Then came La Pérouse, the French commander. He put the fort to the torch, burned everything, and looted a fortune in furs. The French plundered nearly eight thousand beaver and four thousand marten skins from the fort." Prince shook his head as if in disbelief at the loss of precious fur. "Hearne," he continued, "with only thirty-nine men, had little choice but to surrender everything without a fight. La Pérouse had two ships and four hundred men. Still, we were at war with France then, and some thought Hearne should have fought, and died if necessary, to defend his King, country, and the fort. At least the French were kind to him, and took him and all his men as prisoners to England where they were ransomed. But since being sent back to Churchill last fall, the factor has been forced to live in the shadow of his defeat. These ruins are now a bitter and relentless reminder to him.

"He lost his new wife, Mary Norton, too," Prince added sombrely. "He truly loved that half-breed girl. She was left behind with the Chipewyan when the HBC men were taken prisoner. Mary was the daughter of Moses Norton, the previous factor here at Prince of Wales. She had been raised at the fort and was not used to the hard ways of the Chipewyan. They left her to starve. He never forgave them and seems to blame all Indians.

"If any Indian could have changed his mind it would have been old Matonabbee. He was chief of all the northern Chipewyan tribes and Hearne's most trusted ally. Matonabbee guided Hearne on his explorations and organized all the Chipewyan trade for the factor. But Matonabbee hanged himself when the French took Hearne and his men. The chief thought his white friend had been taken out to the Bay and drowned by the French. The old chief had allied himself so closely with Hearne and the British that when the British were believed defeated and all dead, the chief had no more authority over his people, and he was shunned. All his family, his wives and children, having lost their status, starved during that same cruel winter of 1783.

"Now our factor is losing himself in brandy. He's been holed up in that den built from English boards most of the time. I think he's slowly losing his grip. Since he's come back he's done little to rebuild company trade with the Chipewyan, who are selling more and more furs to our competition from Montreal. The Board of Governors in London knows of it and will soon remove him. Too bad – he was one hell of a fur trader in his day."

By October, the river was frozen over and there was little to do but wait out the long winter. The wood they collected was enough for only two fires a day, one in the morning and another in the evening. Most of the time the bunkhouse was unheated. By November, ten centimetres of ice coated the inside walls. On the coldest days David wore a large beaver coat and paced the floor in a vain attempt to stay warm. During mild spells he slipped on his snowshoes and followed Hodges and Prince as they trudged over two-metre snowdrifts and hunted ptarmigan. Most often it was so cold their fingers would nearly freeze before they had time to pull the trigger.

Spring, when it finally came, brought little improvement. With the warm weather, a plague of mosquitoes and blackflies descended. Clouds of insects swarmed around any warm-bodied thing. The tormented HBC men smeared oil and tar on their exposed skin, but no amount of oil, smoke, or clothing could keep the bugs off. Even the wildlife, stung and bitten incessantly, were desperate. Some animals were unable to eat due to the constant harassment. Caribou and foxes alike swam into the river as they searched for even temporary relief. By midsummer the swarming insects had abated, and the apprentice was again occupied with company business, although trading was much reduced from previous years.

David opened his eyes to the bright cold light of late morning. Dew still glistened on the single blanket covering him. He sat up from the patch of ground and knotted roots that had tormented his sleep. Two packet Indians, the name given to Aboriginal People who carried dispatches between the HBC forts, lay near him, asleep. The part-empty gallon of grog that had kept the two men staggering around the campfire late into the drunken night lay beside them. These two were his guides. For the next ten days their job was to take David from Churchill to York Factory, 240 kilometres on foot along the shore of Hudson Bay.

David had been given the barest of provisions as he was told to follow his two Chipewyan guides. The factor's instructions were clear on how his Grey Coat apprentice was to be relocated. While David's provisions for the journey were meagre, Hearne's men had supplied an overly generous amount of grog to the packets. The effect was predictable. The journey would take days longer than needed. But it was the end of the summer of 1785, and David was now fifteen. Nourished in the past months by an abundance of fresh game and clean air, David had fortunately developed a sturdy build. If he had been sent a year ago, when he first arrived from London, his chances of surviving such a journey would have been far from certain.

He waited hungrily for his guides to rouse themselves and make breakfast. He had no food of his own in his small pack. On hearing their charge get up, the heavy-eyed men stuffed the blankets and the bottle of grog into a pack. They half-heartedly motioned for David to follow, and started on a slow, steady walk. There was no breakfast and no lunch either.

The small party climbed over drift logs, waded through tall shore grasses, and skirted their way around marsh pools and rivulets as they struggled across the vast salt marshes of the coastline. Shorebirds and waterfowl took noisy flight, and in the distance, great white bears turned their heads to watch the intruders pass.

Just before sunset the guides set their packs down on a dry stretch of sand near a meandering stream and wandered off, leaving David with the gear. Within minutes the air erupted with the sound of gunfire and the guides returned, dangling three ducks and a goose. Working with a steady and quiet efficiency the packets piled dry driftwood, and a fire was soon sending its sparks into the darkening sky. The birds were plucked and cleaned. The ducks, roasting on sticks, dripped fat into the sputtering flames. The goose was set sizzling into the fire's roasting embers. Now in full darkness, each man gnawed hungrily until the bones were picked clean. The rich oily meat satisfied David's daylong hunger, and it was enough to fuel his body and warm him for the night as he drifted into a deep sleep on the sand.

In the following days the routine continued – long days of slogging along the wet mud near the high-water line. When the mud ended, the ground became uneven. Tufted hummocks and waist-deep channels slowed their march. Seaward, the flats extended to a horizon so distant the sea was hardly visible. Large boulders studded the flats like mute monuments marking the halfway point between low tide and high water. Patches of glistening drift ice still remained trapped among the boulders. Inland, a boggy moor, pockmarked with small ponds, stretched for kilometres before the stunted trees of the sparse woodlands began.

During these days they often passed twelve to fifteen bears a day.

"Don't look at them," cautioned the guides.

David was told to keep a steady pace, as if he didn't notice the bears. When he did sneak a glance, he could see the great white beasts lift their heads to look at them. Often the bears were resting on the flats, with four or five lying in a circle formation with their heads facing the centre.

Avoiding these large predators was not always possible and eventually they came across a bear that was blocking the trail at a stream crossing. It had killed a beluga whale and dragged the carcass onto the bank. They waded in to cross, but the bear gave a threatening growl and showed its fearsome teeth. It would protect its kill from any threat. They moved upstream and crossed at a safer distance.

Several days out of York Factory the guides, now out of grog, decided to take a polar bear hide in for trade. It would fetch them three pints of brandy. David sat on a drift log and watched. Their target, a sow some distance out on the flats, seemed unconcerned by the approaching hunters. When she finally resolved to move away, it was too late. David saw puffs of smoke and heard shots as the bear staggered under the musket fire. The brave animal twisted and turned defiantly towards the shots as the lead musket balls entered her body. When she finally fell, the hunters fired a last shot into her head. That done, they unsheathed their knives and began the long skinning process.

By the time the hide was removed, the tide on the flats had risen to knee level and was soon threatening to engulf the hunters in the frigid water. They quickly severed the head and, dragging it by the ears, waded to shore, leaving the floating hide behind. They placed the head facing seaward on a grassy hummock, and rubbed its nose with red dye. Then the two packets chanted. All of this, they told David, was meant to please the spirit, Manito, so he would cause the hide to drift ashore and not be lost. But it didn't work. *I guess the Manito wants them sober*, David reasoned.

Library and Archives Canada /C-010502/William Armstrong.

Buffalo meat dried in the sun was made into pemmican,
the staple food that fuelled the fur trade.

4

South Branch House

An hour before sunset, the rain finally stopped. Daylight was slipping fast from the river as the sodden and weary brigade beached their canoes. The men found a flat place on the undulating granite that stretched from the water's edge to the dark impenetrable wall of trees above the river. Here on the flat rock the canoes were turned belly-up, laid atop cargo bales, and arranged into a shelter for the night. Someone lit a fire using pitch-wood that had been carefully packed away and kept dry for the purpose. Soon a plume of billowing grey smoke arose like a twisted pillar high above the dank and mouldering forest.

Before long, Mitchel Oman, one of the brigade leaders, handed David a bowl of warm oatmeal and bacon. They both sat in the flickering light of the open fire, and David gratefully leaned his back against a bale and cradled the bowl in his lap. He was hungry, but fatigue overwhelmed him and he fell asleep before he could lift the first spoonful to his mouth. Oman, knowing the boy would need all his food to continue their journey, rescued David's bowl and tucked it under a bale for the morning.

A month ago, in late July, when York Factory's chief factor, Humphrey Marten, had told David he was assigned to the fur brigade going inland, the boy had been unable to stifle a joyful "hurrah!" The news was welcome after a monotonous winter and uneventful spring. David's routine after coming to York Factory was the same as it had been at Churchill, long and tedious hours of clerical work interrupted only by firewood duties. In fact, York Factory and Churchill weren't much different on any account. During the war with France, both factories had been sacked by La Pérouse in 1782 and later rebuilt. Both were major trading ports busy loading and offloading HBC ships.

Humphrey Marten, like Samuel Hearne, was a brutish factor with little patience for a green apprentice like Thompson. But Marten, at least, had reestablished good trading levels after the war. Still, the company's trade was down. More than once, David heard Marten curse his competitors from Montreal as the gout-ridden factor hobbled about the post shouting orders. The "damned, cursed, bloody, peddlers!" had taken advantage of HBC setbacks from the war and established lucrative trading posts inland. The "loathsome bastards" had begun intercepting Cree and

Chipewyan with furs bound for Churchill and York while these factories were still being rebuilt from their ashes.

The HBC had two posts inland: Cumberland House, built by Hearne in 1774 when he still had passion for the business, was over sixteen hundred kilometres' travel from Churchill. Construction of Hudson House, some five hundred kilometres farther upstream on the Saskatchewan River, soon followed. But these posts were the company's first reluctant attempts at building new trading depots. For over a century the HBC had relied on well-established trade routes to bring a steady flow of furs to the shores of Hudson Bay. The new posts were not enough, and Humphrey Marten was damned if he would send ships to London with their holds only half filled. He had no choice but to respond aggressively to the challenge from Montreal. In 1786, Marten ordered his trading brigade to build more inland posts and look for locations to recapture trade from the "peddlers."

David was part of a three-man crew assigned to one of the canoes in the brigade. In all, twelve large canoes were assembled for the expedition. Each was loaded with six forty-kilogram bundles of trade goods. Loading was supervised by William Tomison, the brigade leader. He was making sure each bundle was stowed and lashed so the canoes lay safely in the water.

"Davie, get ye t' the stores," Tomison shouted in a rough Scottish brogue as he tugged a lashing line tight. "Get a good leather coat, a hat, stout mittens, and ye'll need snowshoes too. These won't cost ye laddie, but anything else ye might decide ye need comes out of your apprentice allowance." Tomison then turned to Mitchel Oman. "Take young Thompson in your canoe Mitchel. He'll be your trading clerk for this trip."

Oman nodded dutifully, despite his concerns that he had a city lad from London and not one of his own countrymen to train. The middle-aged Oman had trained many recruits. He was a reliable veteran and, like Tomison and most others with the London-based company, he was not English but Scottish and from the Orkneys.

The Orkneys, a cluster of barren Islands off the north coast of Scotland, supplied three out of every four men working the fur trade for the HBC. They were recruited at Stromness, the Orkney port where David, on the *Prince Rupert*, had stopped. Tucked out of the stormy North Atlantic weather, Stromness was the last provisioning port for HBC and other ships destined for the New World. From Stromness, ships could follow winds due west along the sixty degrees north latitude to Greenland. From there, they went through Davis Strait and into Hudson Bay.

Captains and commanders had long used Stromness to load supplies and the much-valued water from Login's famous well. Henry Hudson had taken on the well's sweet water more than a century before David Thompson, when the renowned navigator first discovered the great bay named after him. Captain James Cook filled his ship's barrels from the well. Sir John Franklin used the Logins' well water, almost a century after Thompson, when Franklin's ill-fated expedition went in search of the Northwest Passage through the Arctic ice.

But the most valuable cargo taken on at Stromness was not provisions or even Login's water but the small, tough island men. These hardy souls, accustomed to eking out a threadbare life on a rugged and unforgiving landscape, made ideal recruits for the harsh demands of the New World.

ꟹ

David helped load the last of the provisions, thirty kilograms of oatmeal, ten kilograms of flour, and fourteen kilograms of bacon, into his canoe. He scrambled aboard and settled into a middle seat between Oman and the bow as they shoved off. The flotilla of canoes, with Union Jacks and HBC flags fluttering off their stern posts, caught the wind and the rising tide, which carried them on an easy paddle six kilometres up the Hayes River delta. This was the last of the easy paddling. The river became shallow, and David had to take his turn hauling the heavy canoes up the rapids by tracking lines. Oman ordered him into the knee-deep water. David slung the thick sisal rope over his shoulder and leaned into it, hauling the 450-kilogram payload by its bow. They pressed slowly upstream. Oman, gripping the gunnels, waded behind and pushed steadily. A second line was strung to a crewman ashore. He was the safety, who helped guide their fragile craft through the rocks.

The river was low and the shoreline trail normally used for tracking was far from the water. The men were forced to wade over algae-slickened rocks. From watching Chipewyan guides, David learned to step between the submerged rocks, thereby avoiding the painful stumbles that came from trying to balance on the rocks' slimy surface. Still, he fell often, and the tracking line slipped loose. When this happened, the second line took up the slack, without which the bark-hulled craft would be quickly swept downstream and smashed. Losing their canoe would mean the three-man crew would have to walk back to York Factory empty handed.

Tracking was backbreaking work in midsummer heat. Swarming mosquitoes attacked their exposed, sweating skin. They rotated duties, alternately tracking, pushing, and lining upstream. David was thankful the river's cold water numbed his aching and rock-bruised feet. The slow ascent went uninterrupted for the next seven days until they finally reached a haulout. The cargo was unloaded and, bale by heavy bale, portaged over a root-strewn trail to the shore of a small lake.

David hoisted a forty-kilogram bale on his back. He slung a trump line over his forehead to help support the load and staggered up to the trail. Then he staggered back for another bale and finally helped shoulder the canoe for the last trip. The heavy tracking was over, but ahead lay nearly two hundred kilometres of lakes, portages, and slow-moving streams, until the divide. Then they would at last travel downstream and westward to Lake Winnipeg.

From David's stumbling gait, Oman knew the lad was in need of several days' rest. He had earned it, but there were no days for rest. David would have to endure the punishing workload until his body hardened into the lean toughness required for the trade.

He might not be an Orkneyman, but Oman had to admire the boy's dogged determination.

ꝏ

The 160-kilometre paddle along the shores of Lake Winnipeg offered no release from hardship for the young recruit. He exchanged an aching back for painful arms and blistered hands. David almost looked forward to the next portage, but his anticipation of relief soon faded. The steep three-kilometre portage took three days of hard labour before the brigade reached Cedar Lake, the last large body of water they would cross before entering the Saskatchewan River. The Saskatchewan was their highway across the western plains, and just a week upstream the river brought them to Cumberland House. They had travelled over eleven hundred kilometres from York Factory. They stopped just long enough to take on a few dried provisions. Tomison wanted to press westward. It was already late August.

Even though only a month had passed, to David, York Factory and his clerical life seemed a world away. The dark conifers and the bug-ridden shield had gradually opened into an expanse of shimmering aspens and sweet-smelling poplar. On the river's bank, ash saplings pushed through tall grasses to reach toward an expansive blue sky. They were averaging forty kilometres a day, and in the evenings, David now had enough energy to rustle up boughs for a soft bed. On these gentle summer nights he heard the haunting bugle of elk and the lulling flow of the river. He wrapped the sounds of the wilderness around him and drifted into deep sleep. This new wild place was giving him strength and confidence and each passing night seduced him further. He never wanted to go back.

The brigade split up at the south branch of the Saskatchewan. Tomison and most of the canoes continued up the main branch to their wintering place at Hudson House. David and the remaining men followed Oman. They took four canoes and tracked up the south branch. On the third day they discovered log buildings under construction on the bank above them. Men, wearing long red and blue caps that hung halfway down their heads, were busy with axes and spokeshaves. Their grey blanket coats were drawn tight around the middle with wide leather belts.

"Canadians" muttered Oman as he steered close to shore to meet them. They were all French speaking except for one they called Thorburn. Oman and Thorburn exchanged courtesies, but the tension between them was apparent. Thorburn told them the Canadians were trading for two different companies under the management of Simon McTavish from Montreal. The next day, Oman, following Tomison's instructions should he come across the "peddlers from Montreal," started construction of an HBC post just slightly upstream. "Build on top of them if you have to!" Tomison had ordered. Oman named the new post South Branch House.

Trading with the local Cree began before the posts were complete. David enjoyed the Crees' company and listened attentively when they spoke. He began to pick up more than the simple trade words Oman had taught him along the way. In time, he was able to communicate easily with them. He found the Cree to be proud and independent, unlike the Chipewyan at the Factory. These men were tall and fine looking in leather clothes painted with red and yellow dyes. Some were draped in shining buffalo robes. They had high cheekbones, prominent noses, and dark eyes. These features emphasized what David saw as a dignified and self-reliant bearing. The women dressed much the same as the men and were fond of the bright cloth David sold. The Cree had only a few horses, which they kept for hunting and bringing meat to their tents. Dogs were fitted with side bags to carry belongings or harnessed to long poles for hauling buffalo or elk hides to camp.

The Cree found their beaver and wolf skins could be sold for prized sewing awls and strong needles to replace the troublesome thorns the women used for sewing leather. Flint and steel could be used to make fire on demand. Now the old Cree, instead of anxiously guarding glowing embers in a wooden bowl and constantly feeding sticks to the coals, could move fire easily. Most of all, guns and ammunition could be bought. Victory over an enemy tribe now depended more on the number of guns than on the number or the bravery of warriors. To the Cree, it was unbelievable that white men would trade such valuable items for only a few animal skins.

The Canadians also took their quantity of furs. In fact, they were out-trading the HBC post. Although the goods from Montreal were inferior, they had brought four times as much as the HBC brigade had. The

Canadians had brought many iron pots and cloth coats. Oman had brought copper pots and heavy wool blankets, which he knew the Cree preferred. Most notably, one-quarter of all the Canadians' trade goods were barrels of strong liquor. Drunk straight, it was too strong and was like poison to the Aboriginal Peoples, but when diluted with water, they found it irresistible.

By the beginning of April, trade with the Cree was over. They had followed the buffalo to the open plains. As David and the others waited for the river ice to break, they sorted furs and pressed them into bundles. The Canadians had done very well. At break-up, they paddled their loaded canoes toward Lake Superior and Montreal. The HBC brigade paddled in another direction. They set out for Cumberland House, where they would rendezvous with Tomison for the return trek to York Factory.

For David, returning to York Factory would be a sentence to months of boredom. Oman knew this and put in a word with Tomison. When Tomison's canoes left for York Factory he let David stay behind at Cumberland House with two HBC men.

"You'll spend the summer here, Davie me lad. Barter the remainder of the trade goods and help set in fish and game provisions. We'll need food for the gang returning this fall," instructed Tomison.

The summer passed quickly enough. David tended three long gill nets strung out on Cumberland Lake. The fish were smoked or rendered for lamp oil by the men at camp. Cree hunters supplied all the meat they could want. Their women hung out strips of fresh red meat to dry in the sun. David helped pound dried buffalo meat into powder, which the Cree women then mixed with fat and berries to make pemmican, the staple food for the trail. Pemmican fuelled the brigades far better than oatmeal and bacon did. Less than a kilogram would provide a man all the energy and nutrients he needed for a day's travel. In years to come, a wilderness industry developed to supply pemmican for the fur trade. Plains communities of mixed French, First Nations, and Scottish ancestry, called the Métis, would emerge and thrive on pemmican production. Each year Métis hunters killed and quartered buffalo by the thousands. They dragged the meat and hides to camp where the rest of the community efficiently processed the animals into pemmican bricks for sale, barter, or trade.

In late August of 1787, Tomison returned to Cumberland House with another brigade and canoes packed with goods. Some of his men were in poor health or lame, and they were left at Cumberland House. David joined the remainder of Tomison's party, heading farther inland to Manchester House, the company's new post 160 kilometres up the Saskatchewan River. They journeyed west through gently rolling hills and crossed the paths of never-ending herds of buffalo. Wave after wave of the huge beasts swam across the river, striving for fresh grasslands on the other side. More than once, Tomison and his men were forced to beat the animals with paddles to fend them off and avoid being swamped.

The canoes pushed upstream to the steady rhythm of paddles until they reached Manchester House. The post was simply a rough log building on the riverbank. In front, a lone figure in full native dress stood squarely with feet apart and both hands on his hips. James Gady, bearded and stocky, looked as if he not only owned this post but everything west of Lake Winnipeg. Gady had spent two years in this country living with the Peigan. He had learned their language, customs, and ways of survival. He could sustain himself in these grasslands without the support of either Peigan or white man, and because of this, he was well respected. He was a friend of the Peigan and a valuable asset to Tomison and the HBC.

"It's about time you sore-assed tenderfoots got here," bellowed Gady, as the first of the canoes gently buried its keel into the sandy beach. *He's right*, David thought, *my ass is sore. Fourteen-hour days of paddling on a hard canoe seat will do that.* Like drunks leaving a tavern, Tomison and his men staggered from their canoes until their legs regained circulation. Tomison shook Gady's hand as David and the crew busied themselves hauling cargo up the shore to the grassy bench by the house.

That night, by lamplight, Tomison and Gady prepared a plan for an overland trading expedition deep into Peigan country. A party under Gady would take horses and enough trade goods to secure positive relations with the Peigan. David overheard his name spoken as a possible member of the party. Gady lifted the lamp to see the young recruit in better light. He looked Thompson over with a critical eye then nodded his okay to Tomison as they resumed the discussion of details. David swelled with pride and excitement. In the last two years he had gone from a pale company clerk to become a member of one the HBC's farthest reaching explorations.

Gady took six men. Each were loaned trade goods and two horses at a discount and expected to repay their debt and make any personal profit from the furs they would sell to the Company. They rode out in the first days of September. David followed behind Gady on foot. As the newest fur trader, he had only enough borrowing power for one horse, and it was fully loaded. He was thankful Tomison had given him a new leather coat and a buffalo robe as well as ammunition and two long knives to help him on his way.

"The Peigan," Gady told David as he jogged beside the leader's mount, "will be different from other Indians you've known. They won't be found begging for handouts around the factory walls. These are Plains Indians. Like the Blackfoot and Blood, they're warriors. They raid other tribes and fight battles on open ground. You need not fear having your throat slit while you sleep. They're too proud for that. If they take a mind to kill you, they'll do it face to face. And remember, they don't need us except for the guns and the few luxuries we trade. They can travel across these plains free and easy, following the buffalo for thousands of miles, and they find all that they need or want."

David trotted beside Gady day and night to learn about the Peigan and their language as the fur traders entered the magnificent homeland of the Plains people.

Library and Archives Canada /C-010513/William Armstrong.

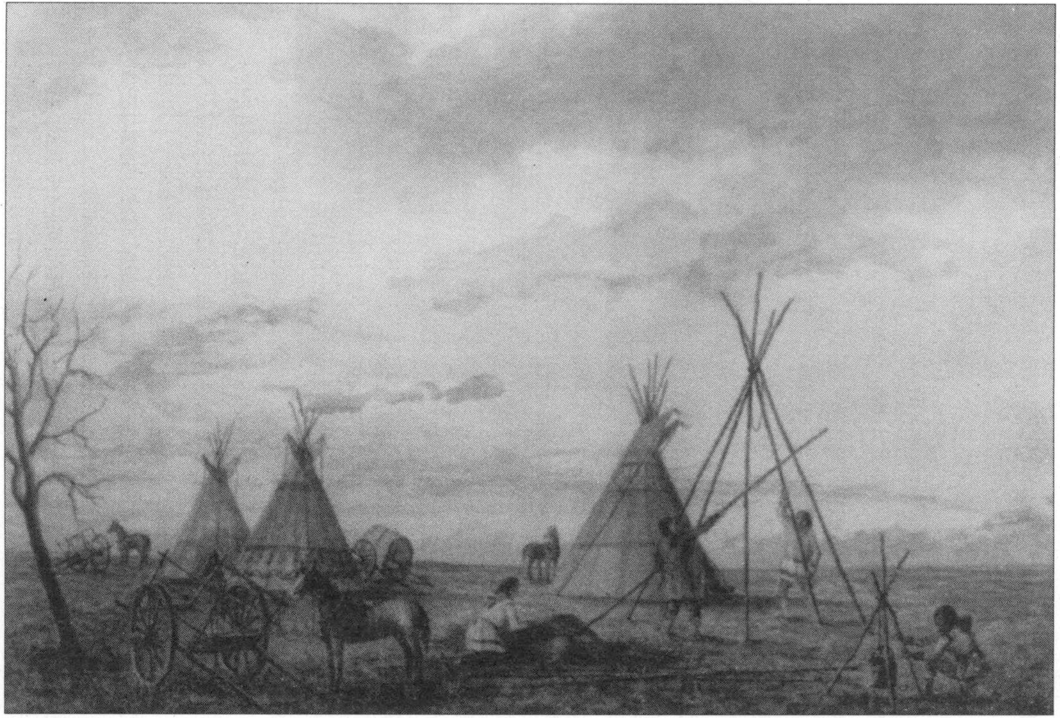

Living with and befriending the Aboriginal Peoples of the Plains gave Thompson an advantage over other explorers like the Americans Lewis and Clark.

5

The Plains

A Peigan war chief knelt beside an earthen pit already dug atop a dry, grassy knoll, not far from Gady's camp. His name was Kootanae Appee, meaning Kootenay Man. He had a high forehead, an aquiline nose, and dark eyes that defined a face both honest and fierce. He stood nearly six and a half feet tall. Lean and agile, he was built for fighting. But on this day his thoughts were far from fighting. He was seeking solitude to reconnect with his spirit. As was his custom, he had ridden far from his village and climbed a distant hill. Here, he would catch a hawk whose feathers would adorn his new headdress.

Using the age-old method, he lay face up in a pit dug like a shallow grave. He reached out for the stack of branches and laid them across the opening. Then he pulled dry grasses and leaves atop the branches, completely concealing his presence. Lastly, he placed a freshly killed rabbit, impaled on a stick, on top of the leaves and rested it just above his chest. When the shadow of the hawk appeared, Kotanae Appee would be very still. When the rabbit moved, he would reach out and grab the great one by both legs then beat the struggling predator against the ground until its vicious beak and talons were no threat. Meditating and softly chanting, he would lie patiently for many hours, until his feathered quarry descended.

There were few trees about, so David searched the grass for dried buffalo dung in place of firewood for their camp. He followed a trail of droppings to the base of a grassy knoll.

Above, a hawk circled excitedly in the blue sky, then descended, wings tucked, into a dive. David recognized the behaviour. He had seen these airborne predators hunting the prairie grasses for small game many times before. He waited expectantly for the hawk to sweep up again, carrying some luckless gopher or rabbit in its talons, but strangely, the bird never reappeared. David meandered back to camp, stooping occasionally to add another dried chip to his armload.

They had travelled twenty-four kilometres a day, giving the horses plenty of rest and time to graze the nutrient-rich grasses. They hunted, but game was unusually scarce and the men were hungry. Gady could not explain why there were no buffalo or elk.

Within a month the small trading party could see the glimmering white tips of the distant Rocky Mountain Range. As they trekked closer, the mountains rose into immense snow-topped masses, piercing the clouds and forming what David thought must be an impassable wall. A few kilometres beyond the Bow River they met the first Peigan people. Unknown to Gady or his men, scouts had been watching the traders for several days before they finally approached with a dozen warriors on horseback with their quivers full of arrows.

Fortunately, their welcome was friendly. They told Gady to camp where he stood and that they would return with fresh meat. "We must look a sight," said Gady. "They know we haven't eaten anything worth spit in weeks." That evening the Peigan and Hudson's Bay men feasted on fat cow buffalo and talked well into the night. The Peigan wanted news of other tribes – their numbers, was there disease? Were they at war? Gady told them all he knew, which wasn't much.

In the following days the traders were taken to the Peigan's main camp, where they traded ammunition and tobacco for choice fox and beaver skins. David was lodged in the tent of a grey-haired elder named Saukamappee. The old man was solemn, but mild-spoken. His tall frame was still strong and remarkably supple. Strangely, Saukamappee was pleased to hear the Hudson's Bay men had found no game. "That means a plentiful winter," he reasoned, "because the buffalo have stayed too long on the Missouri grass and will be hungry for new grass near our mountains."

The old man was pleased even more when he found David could speak the language of the Cree. He asked David about his father and mother and their country. David explained he never knew his father, and that his mother had given him away to a school where he and other boys were raised to help navigate ships that crossed the great oceans, but he had instead been sent here to trade furs.

The old man nodded. "I am not a Peigan," he said, smiling after a long silence, "but a Cree of the Pasquiaw River near what you call Cumberland House. I too, came here as a young man. I came with my people to help our friends, the Peigan, in war. My name, Saukamappee, means 'young man.' I have not returned to the country of my people since. It warms me to speak with you in the tongue of my mother, which I had almost forgotten."

David told the old man what he knew about the Cree near Cumberland House. He mentioned the names of some of the elders there, but Saukamappee knew none of them. "I see I am now a stranger in the land of my father," he said, and then began to tell David about his years with the Peigan. The old man was fond of speaking in his native language. Nearly every evening he related the events of his life to David.

"All these plains are now shared by three tribes," he told his eager young audience. "The Peigan, the Blood, and the Blackfoot. We speak the same language and help each other in war. But it was not always so. In the past, this country belonged to the Kootenay in the north, to the Salish in the west, and to the Shoshoni and their allies in the south. They were many more than us, and they were our enemies, especially the Shoshoni, who would attack the small hunting camps of the Peigan, killing everyone. Then the Peigan sent messengers to my people asking for help, so I came with my father and twenty other Cree warriors to the Peigan. I was proud to be included in the great war tent when we feasted and made war speeches and danced. The Peigan were much pleased that we had guns from the white men.

"One day our scouts told us they had seen a large camp of the Shoshoni not far from us. They were many." Saukamapee opened and closed the fingers of both hands, then held up one finger signifying that each finger was worth ten. He then slowly and sombrely stretched open both hands three times and one hand once. *Three hundred and fifty*. David nodded and waited intently for the old man to continue.

"When we went to fight them," Saukamappee said gravely, "our war party stood opposite facing the enemy. The enemy danced and shouted war cries and then they crouched behind their large shields. We did the same, but we had fewer shields, so two warriors knelt behind one shield. Then the arrows came. Their bows were not as long as ours but were made of better wood, and their arrows shot between us and sometimes went right through our shields, wounding us. Our arrows did not pierce their shields, but stuck in them.

"We became afraid that we would lose and had fear they might bring horse warriors. Back then my people had never seen a horse but had heard how the Shoshoni could ride them swift as deer and how they used stone clubs to smash the heads of many warriors who could not run from them. But we were fortunate because they had no horses, and they had not yet traded with the white man so the Shoshoni and the Kootenay did not have guns.

"Our war chief told those of us with guns, of which I was one, to lay between our shields. We had ten guns and each of us had thirty balls and powder. The chief told us to shoot, but we said we must be closer, so we moved to where their arrows could easily kill us. But when a Shoshoni warrior stepped from behind his shield to fire his arrows, in the way they always do, we fired, and our bullets never missed.

"We killed and wounded many that way until they began to crawl away and hide from the guns. Our chief signalled and we ran after them, killing them before they could run away. We now wanted scalps to honour our battle. Some brave Shoshoni stayed and fought but there were five or six of us, all trying to get the scalp of one of them, and it was soon over and only a few scalps had been taken., Then some of our warriors began scalping the dead ones, but these scalps were poor trophies and had no value to me. Taking a scalp captures the soul of the enemy, but only for the one who kills him. So I took no scalps because I did not know which ones I had killed with my gun. I was unhappy because my wife's father wanted scalps to honour his medicine bag and to give to his ancestors.

"After the victory feast the chief asked all the warriors to give their scalps as gifts to families who had lost relatives in the battle. Our people believed the souls of the slain enemies would then become slaves for our dead ones. But many warriors did not know what to do because the chief would not let scalps taken from the dead dishonour the families. In the end, he said they must give these scalps to us, the Cree, who had killed so many with guns and who had won the battle. I was given many scalps.

"After that I dressed in my finest leather and painted my face and body so that I would impress my wife and her father when I gave them my scalps. I rode many days to find them, but when I came to my people's village they said my wife was no longer with them. She had taken another man before I was three moons gone and moved to another village with her family. My heart was swollen with anger and revenge but my Peigan friends said she was worthless and I would find a better wife with them. That was when I renounced my people and returned to live with the Peigan. The Peigan chief then honoured me with his eldest daughter for a wife. She has now grown old with me.

"The terror of that battle and our guns stopped the Shoshoni and Kootenay from attacking us. But we did not stop. We warred for many seasons and we killed as many of our enemies as we could find. Finally, we drove the evil ones, the Shoshoni, the Kootenay, and the Salish out of this land and across the mountains." The old man paused and seemed to fight back tears as he summoned strength back to his voice. "But the killing brought an evil spirit to us, which entered the bodies of men, women, and children, killing everyone and destroying whole villages."

Smallpox! David thought.

"After the great death everything changed," said Saukamappee. "Now there were not enough people for our village. We believed the Great Spirit had forsaken us and was angry at the blood we had spilled on his land. So after that we killed only the men. The women and young ones were taken back to our village to live with us. Then our villages grew strong again. That is how we got this land."

In mid-winter, a small war party of young men returned to camp after a long absence. They sang a war song of victory, and one came to the old man's tent to pay his respects. After he left, Saukamappee told David the warriors had been gone for two months, and the famous war chief, Kootanae Appee, would arrive tomorrow with the main war party. They had gone far south to meet their enemy, the Black People (the name they gave the Spanish), and capture their horses. David asked about the battle and the old man smiled. "No battle. The Black People never fight. They always run away."

He told David it was just as well for the Spanish because Kootanae Appee was a brilliant battle strategist, and it was he who had defeated all the other tribes and carried the Peigan to great heights of power. Saukamapee said the war chief was famous for snatching victory while retreating. Unlike war chiefs before him, who left warriors to fend for themselves, Kootanae Appee would organize strategic ambushes. His rearguard actions won him many victories and the devotion of his warriors.

The following day the tent flap swung open and Kootanae Appee entered Saukamapee's lodgings.

There was no ceremony between them. They were old friends and each held high status in his tribe. The old man presented David to the chief, but Kootanae Appee's stone-faced stare gave no hint of friendliness. The war chief extended his left hand in greeting and David gave him his right in return. In doing so David had unknowingly challenged the chief to combat. The chief smiled, for he knew the contest would not be equal. Saukamapee intervened quickly. "Listen young man," he scolded. "If one of our people offers you his left hand, you give him your left hand. The right hand is for fighting. It wields the spear, draws the bow, and pulls the trigger. It is the hand of death. The left hand is next to your heart, speaks truth and friendship, and is the hand of life. When you leave this place you will meet many who have never seen a white man. So remember your manners."

They all sat. Kootanae Appee and Saukamapee talked for an hour before the chief left as unceremoniously as he had arrived. Afterward, Saukamapee told David he had recommended him into the chief's protection and Kootanae Appee had agreed. "This is an honour," he explained. "You have a great warrior's guarantee to protect you in his territory."

David was sorry to finally leave Saukamappee, but he needed to follow Gady and the others on their trading mission deeper into Peigan country. David remembered what the old Cree had taught him, and was able to accumulate a good number of valuable furs trading with the Peigan.

ꝏ

William Tomison raised his arms and smiled warmly as he greeted Gady, David, and the others as they returned to Manchester House. He already had news of their trading success and eagerly examined the bales of fresh pelts. "All prime. You'll no find better than this," he said, sucking air across his teeth and shaking his head in admiration at the quality of their furs. "Eighty skins, Davie. Well done, lad!" congratulated Tomison as he rummaged through David's bales. David could not remember ever seeing Tomison so animated and wondered if the brigade leader was happier about the furs or the men returning.

When the river ice cleared, the brigade, laden with a small fortune in fur, left for the shores of Hudson Bay. David, thankful he had been left at the post until next summer, began helping with the upkeep of Manchester House. He cared for the company's forty horses by giving them a small evening feed of oats and corralling them in for the night. When winter set in, collecting wood to keep the fire burning at the post was an endless duty.

On a bitterly cold morning two days before Christmas, David headed out, dragging a wood-sled behind him in the crusty snow. He trudged past some good fire logs and mentally noted their location. *Get them on the way back*, he thought, *no use hauling them both ways*. Finally, he stopped at a promising pile of fallen trees. They were down a steep bank. Carefully leaning his weight on a tree branch, he stepped over the edge. The branch snapped and he began to slide, almost playfully, until his leg became wedged under a fallen log. His leg was stuck but his momentum continued to propel him forward. He heard a sickening snap and felt searing pain pulse through him as his face hit the snow. He tried to get up, but a rush of pain and nausea overcame him, and he fainted into a nightmare of suffering. His leg was badly broken.

He regained consciousness to an acrid smell and opened his eyes to find a puddle of brown vomit melting the snow by his mouth. David tried to move, slowly pushing with his good leg, until the pain became too great. He waited, then pushed again, gradually working his way up the bank. The others wouldn't notice him gone until dinner. By then it would be dark and too late to find him. David doubted he could survive through the frozen night. He pushed to the edges of pain, then with growing desperation, pushed harder into the very core of his pain, into the red blinding flashes that thundered through him. He pushed until he heard voices, felt hands lifting him, then darkness.

David awoke at Manchester House. The events of the previous day began to take shape in his mind as he found himself lying on a bed near the fire. He pulled back the stale buffalo robe covering his leg and surveyed the bindings that had been wrapped around it. The men at the post had done their best for him when they found him crawling, half conscious, toward the post. They knew vaguely how to bind his broken leg with a bandage and splint, but none had any medical training, and the injury was severe.

For months, David was unable to stand and was still unable to walk by spring thaw. Tomison was angry.

If there had been a surgeon to set young Thompson's leg, he would not have lost one of his best young prospects. The brigade leader went out of his way to speed David's recovery, and any doubts that David had about Tomison's sincerity vanished. The brigade leader showed him the kind of tenderness David thought a father might give a son, but in the end, Tomison was forced to send David downriver to Cumberland House.

Here David fared even worse. He was left in the care of the two men stationed at the now inactive depot. They had little understanding of nursing the ill, and soon David became weak and unable even to feed himself. He was emaciated and nearing death when a kind-hearted Cree woman took him into her care. Using fresh berries and healing herbs, she nursed him back to life. He was thankful to her for his restored health, but he still could not use his leg.

As the darkness of winter came again, David began to face his bleak future. With a bad leg, his only option was the dreary duty of a trading post clerk. He missed old Saukamappee, and he longed to ride back to the open plains and the Bow River and to feel again the prairie wind.

6

Cumberland House

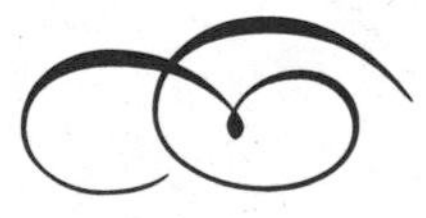

The sun had left the sky for many hours when David limped onto the frozen surface of Pine Island Lake. The ice-crusted hem of his buffalo robe dragged in the snow, leaving a track of brush-marked footprints that led far out from the dim lights of Cumberland House. He wanted to see all of the night sky. He cast the heavy robe fur-up onto the snow and lay upon it, gazing skyward. The stars in the immense blackness captivated him. Some were just faint imaginings, more like speculations than stars. Others, like beacons from a nearby shore, were brilliant and dazzling in the heavens. These were his signposts, especially Polaris, the gleaming North Star, and the two guiding constellations of Cassiopeia and Ursa Major, which flank Polaris

Library and Archives Canada /C-001426/ C.W. Jefferys.

David Thompson Taking an Observation.
His trail by stars. Using a sextant, compass, and the stars in the sky,
David Thompson mapped Canada's vast wilderness.

east and west. "These stars will tell you latitude," Philip Turnor had explained with enthusiasm early that morning.

David could hardly believe his turn of fortune. He had been a trading clerk for only two months and Tomison's brigade had just recently departed when two canoes arrived unexpectedly at the post. Among the arrivals was Philip Turnor, the famed HBC surveyor and prominent man of science. He had helped compile the *Nautical Almanac* used by mariners around the globe for celestial navigation. Part of Turnor's assignment in Rupert's Land was to map the route to Lake Athabasca, but he was also to train a select few Hudson's Bay men as company surveyors. These new surveyors were to meet the growing need for maps of the ever-expanding trade routes. If his leg continued to improve, David could be one of them.

"One of the greatest scientific advancements of our time is being able to describe where we are on the earth's surface. James Cook's voyages are celebrated not so much because of where he went, but because he was able to chart his voyage in latitude and longitude with exceeding accuracy. His charts help Britain dominate the world's oceans," explained Turnor, as the middle-aged scientist and David warmed themselves by the afternoon fire.

"If we wanted to send someone to Montreal, it would not be difficult because people know where Montreal is. It's up the St. Lawrence River, one thousand and seventy-four miles from the river's mouth. Or, it's a two-day paddle east of the last portage on a well-travelled trade route. Closer to town, they could follow signs and roads. But what if, like us now, one sits on the edge of a vast wilderness with no roads, signposts, or maps? We can follow the rivers, which is what you and I have mostly done, or we could hire an Indian guide to take us to the next big water. But how do we tell the world how to retrace our steps? Well, we look to the stars."

David knew all this, but somehow, with Turnor, it was like hearing it for the first time, and it filled him with a hunger to know more.

"The moon travels across the sky like the steady hand of a clock," continued Turnor. "It moves approximately its own width every hour. Behind the moon like numbers on the clock's face, are the stars. When the moon passes a known star, the time is told."

The Astronomer Royal, Sir Nevil Maskelyne, had worked tirelessly at the Royal Observatory in Greenwich compiling tables that predicted the path of the moon and other heavenly bodies. Turnor had assisted him, as both scientists painstakingly tabulated the exact time at Greenwich when the moon passed certain stars. The tables were updated every year with new and more complete information and published as the *Nautical Almanac*. Earlier that day, Turnor had handed David his own first copy.

As David lay in the snow, he observed the moon on its journey, and he knew he would learn to calculate the time at Greenwich by referencing Maskelyne and Turnor's tables. He could find the exact local time by resetting his watch each day by the noon sun. The difference in time between zero longitude at Greenwich, England, and where he lay in the snow, would determine the longitude of Cumberland House. This is because the earth also turns like clockwork, as it rotates around its axis fifteen degrees of longitude for every hour.

It wasn't quite that simple, of course. He thanked his old teacher Thomas Adams for drilling enough trigonometry into him so he could struggle through hours of exacting calculations. Performing the mathematical corrections for refraction and parallax that Turnor demanded was tedious work, but it was worth it. Surveying would be his escape from the company store, especially since his leg was getting stronger each day.

"Latitude is less problematic," Turnor told him, "because there is a relatively painless method," the astronomer added quickly, as his new student paled at the prospect of more calculations. "Latitude can be found by simply measuring the angle of Polaris above the horizon. The North Star is always over the North Pole no matter where in the world it is viewed from. The earth spins and swings like a ball on string suspended from Polaris and is therefore constant among the rush of stars travelling from east to west across the night sky. At dusk or dawn, when the horizon can still be seen in the dim light and Polaris is visible in the murk of the transitional sky, the angle read on the sextant is almost equal to the latitude on which the observer is standing. Some corrections have to be applied to get the exact latitude, but a very convenient method nonetheless."

The next morning, David and the men of the post heaved the last few shovels of frozen earth onto the fresh mound that covered Mr. Hudson's shallow grave behind the storehouse. Hudson, Turnor's assistant, had been sick when their canoes arrived, and he had grown progressively weaker until finally the unknown ailment took the assistant's life. That evening, David returned to his quarters and began practising refraction calculations, partly to forget about Hudson's death. The grave could have easily been his own just a few months earlier. But he also studied to ready himself. Turnor would need a new assistant.

Accidental? David wondered. *A circumstance of birth sent me to Grey Coat where it was predetermined that I study mathematics and navigation. An unavoidable injury suffered in the wilderness fated me to study under a famous astronomer. Now Hudson's death means Mr. Turnor needs an assistant. Maybe I was never supposed to navigate a ship but was meant to be a map-maker and chart unexplored lands.*

The prospect of being able to help map the vast unknown like the famous Captain Cook kept David studying under a flickering candle late into the winter nights. By winter's end David's eyes were inflamed from working under dim light. Still, he drove himself until Turnor made ready to leave Cumberland House for his mapping expedition. By that time David was near blindness, weakened by an infection, and unfit to go. He was forced to stay behind while Peter Fiddler, another young clerk, took his place at Turnor's side.

Before Turnor left, he called David aside. "So, young man," he said, "Do me a favour. Don't try so hard next time. They tell me you're being sent back to York Factory as soon as your eyesight recovers. Your leg has healed nicely so I've sent a letter to the factor recommending you as a surveyor. Keep faith that your time will come. And use this," said Turnor, handing David a brass sextant. "The radius is small but it was made by Peter Dolland, one of the finest makers in England. You can use it 'til we meet at York Factory next year. All the best, Mr. Thompson," the astronomer said, handing David the sextant's storage box.

Turnor's party shoved off, leaving David standing at the lake's rocky shore with his dreams held in that small wooden box. One day, he hoped, the Company might ask him to cross the mountain range to the west. Then, when one of the creek beds he'd been following for days dwindled to nothing and it was the end of some river and the edge of nowhere, he'd make his own map and his own choices. Maybe the company would tell him to push west, but only he could decide whether to first go northward or south. He'd be the one to choose whether to follow some untravelled valley and hope to pick up the beginnings of a new river or perhaps crest some distant ridge into another drainage. Whatever his choices, right or wrong, he could trust the stars and his sextant, and they would help him show others where he had gone.

"June 9th, 1790 – Cumberland House – north latitude 53 degrees, 56 minutes and 44 seconds; west longitude 102 degrees and 13 minutes," David muttered as he pencilled the first entry into his journal. "Wind, sou-sou east, overcast, 54 degrees F." Each day over the coming months on his way back to York Factory,

David entered similar data into his journal. *Joseph Colen, the new factor, will be impressed*, thought David, when the route from Cumberland House to York Factory was completely surveyed for mapping.

Colen, however, was a man of business with little patience for meteorology or map-making. He was under pressure from the North West Company, whose traders had penetrated his territory and were siphoning off valuable HBC furs. Colen's future and chances for promotion in the company depended on how well he was able to compete with the rival company for market share. When Thompson arrived, Colen immediately set him to work in the warehouse overseeing the grading and bundling of furs. When that work was completed, he sent him by the same route back to Cumberland House to collect more furs for the factory.

Even Turnor, when he arrived a year later, could not persuade the ambitious fur merchant to see the value of a map. With Turnor's influence, however, David was appointed surveyor. He was at the end of his seven-year apprenticeship, and the directors in London gave instructions that he was to survey another trade route to the rich Athabasca country. But Colen knew how to sidestep orders from London. He could make sure his newest recruit wasn't wasted on mapping Athabasca, but was used as he needed him – for fur production and the commerce of the factory.

For the next five years Thompson was sent trading into the Muskrat Country, the well-known, waterlogged, bug-ridden region between York Factory and Reindeer Lake. He took astronomical observations and fixed positions as he travelled, but routine company business, piled on by Colen, occupied most of his time. Finally, in 1796, Colen couldn't stall London any longer, and he reluctantly gave Thompson permission to map a new route to the Athabasca. Colen could still influence the outcome, however, and he knew that without good men and sufficient supplies the expedition would likely fail, thus putting an end to London's incessant demands. Colen also had Thompson's most recent editions of the *Nautical Almanac* from London. For the last two years he had somehow forgotten to send them forward. What Colen didn't count on, though, was the determination of the young surveyor.

On June 10, Thompson headed north on the Reindeer River. He was spared no company men and given no canoe. Undeterred, he hired two inexperienced Chipewyan youths as guides and constructed a birchbark canoe himself. The meagre ration allowed him meant his small party would have to rely on a fishnet and a musket for food as they advanced toward the treeless northern barrens. Secured in the belly of his five-metre canoe were his surveying instruments and notebook, a small tent and bedding, thirty rounds of ball ammunition for caribou, two kilograms of shot for geese and ducks, two kilograms of gunpowder, and three spare flints.

They travelled fast under the light load for the first week, but eventually the rivers lost depth and they began tracking. When the waterways shallowed further, they carried their canoe and pack for eighty kilometres over rocks and mosquito-infested marshes before finally reaching Lake Wollaston on June 23. Over the next two nights, while his guides rested, Thompson was busy with sextant and pencil completing notes for the survey.

They paddled across windswept Lake Wollaston for several days before finally descending the treacherous rapids of the Black River to Lake Athabasca. Six hours onto the lake they came to a protected shore well suited for a canoe camp. To their surprise, the site had been used before. Here, on a blazed pine tree, Thompson discovered the survey marking of his old master. It read "Philip Turnor – 1791." Finally, after five years of frustration, David had mapped another route to the fur rich Athabasca region. Now he faced a bigger challenge: the long journey home.

Trail weariness had crept up on the men, depleting their strength imperceptibly each day, until they were exhausted. Such fatigue, unheeded, compounds even small problems. In this weakened state their judgment was impaired and their reactions were slowed. That is how it was when Thompson tried to line his canoe homeward up a stretch of rapids. The young Chipewyan guides, weary and unmindful, allowed the river's current to shear the canoe cross-current. Thompson, slow with his paddle, corrected too late and was now faced with the threat of capsizing. He sprang forward and cut the bowline with his knife, but he had forgotten the falls behind him. By the time the sluggish crew noticed the line was slack, Thompson was plunging headlong nearly four metres over the cataract. He was battered brutally against the rocks below the falls but managed to surface, grasp the canoe, and swim to shore. He lay gasping, his body bruised and the flesh torn from the heel of his left foot, while what few provisions they had were carried off and lost in the river.

The crew later recovered the cork-lined box containing Thompson's instruments and notebook but they could not find the fishnet and ammunition. Once Thompson's foot was bandaged with tent canvas and the canoe was repaired with spruce gum, they struck out again for home. The men knew that with only berries for food, they would soon die. In desperation, one of the Chipewyan stole two young eaglets from a treetop nest. The fledglings put up a fearsome struggle and sank their talons deep into the young man's arm before he could kill them. The men ate the birds' yellow fat and saved the meat for future rations. The fat, however, was contaminated and produced severe gastroenteritis. Thompson and his men spent the night and the next few days disabled by vomiting, diarrhea, and intestinal cramps. Weak, and with no extra clothes or fire to warm them, they could not continue. Death seemed inevitable as the cold of the night closed in.

Fortunately, in the morning a passing band of Chipewyan discovered the near-lifeless men and fed them some nourishing broth. In a few days the Chipewyan moved on, but they left behind a small cache that included powder and ammunition. The men, now recovered, could hunt for food.

Some weeks later Thompson and his crew arrived at Fairford House, weak and emaciated but alive. By August, a new brigade had arrived with trade goods, and over the winter, Thompson persuaded the brigade to accompany him on his new route to the Athabasca. This time he had a party of seventeen experienced men, four canoes, and plenty of supplies. Still, the outcome was almost as disastrous as the first trip. The route, although shorter than Turnor's route, was not practical for the fur trade. The waterways were often too shallow and the eighty-kilometre portage made the movement of heavy canoe loads overwhelming.

The expedition was forced to a standstill late in the season and had to winter in a makeshift log hut, which they named Bedford House. The winter was hard. Game was scarce, temperatures plummeted to record lows, and worse still, trade for fur was unprofitable. When word reached York Factory, Joseph Colen finally had what he wanted. Thompson's mapping expedition had failed to produce commercially useful results, and the young surveyor was ordered to return to the business of trading full time.

David Thompson was now twenty-seven years old, and he was angry. His second term of service with the company was up, and until this point he had done everything they had asked of him. His years of hard travel had made him as lean and tough as any veteran fur trader. He was a crack shot and a fine hunter. He could speak the language of the Chipewyan, Cree, and Peigan and they respected him. He could not only read and write at a time when most men could do neither, but he had also mastered the mathematics and astronomy needed to map the wilderness. He felt that for the first time in his life he was free to choose his future, and he knew he had become a valuable commodity.

Yes, the company had offered to promote him to chief trader with the title "Master to the Northward," but to him this was a sentence to years of mundane administrative duties. It would mean overseeing staff, reviewing budgets, and working for higher productivity. Who would care how many furs the "Master to the Northward" had traded in his career? Would it matter to anyone but a handful of wealthy London shareholders how much he increased the company's profit? He wanted his contribution to be more substantial than that. He wanted to map the new wilderness, to contribute something that would endure and benefit many.

He had all his savings in the Hudson's Bay Company bank set aside, with instructions to send regular payments to his mother in England. He sent a letter to Joseph Colen asking that his *Nautical Almanac* and survey instruments be forwarded. That done, on May 23, 1797, Thompson set out for the nearest North West Company (NWC) post at Fraser House, 120 kilometres away. The North West Company needed an explorer and surveyor. Now they would have one.

Library and Archives Canada/C-082974/H.A. Ogden.

At the Portage.

Over and over again. Each man packed heavy forty-kilogram loads strapped to his back as the canoes were transported over land from one waterway to the next.

7

Grand Portage

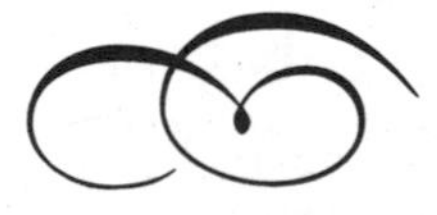

Thompson paused to rest as he completed the fourteenth and last kilometre of the long carry descending into Grand Portage. *Grand Portage…* David pondered the place name as he adjusted his pack a little. *So this is the centre of the Northwesters' fur trade. For good reason too*, he concluded. The strenuous fourteen-kilometre portage connected the North West Company headquarters and the high point of the land. *A better location could not have been chosen*, he mused. David knew that from the strategic high point, three major drainages radiated outward to three distant and significant bodies of water. The Saskatchewan drains into Hudson Bay, the Mackenzie flows to the Arctic Ocean, and the St. Lawrence courses to the

Atlantic. All NWC traffic to and from the vast fur regions went through Grand Portage on its way to Montreal.

There was no mistaking the headquarters of the NWC. Thompson could smell the smoke from its cooking fires over a kilometre away. As he trudged closer, he could hear singing, shouting, then a chorus of barking dogs. Finally the stench of canine and human waste greeted him. It was the largest and most unruly settlement David Thompson had ever seen. Outside the towering five-and-a-half-metre stockade, a haphazard arrangement of birchbark lodgings housed Aboriginal People from several tribes. Scattered among these were a few ramshackle log huts belonging to the independent traders. They were busy skimming what trade they could from the Chipewyan, the Cree, from incoming parties, or from anyone discontented with the Company.

Across the stream, a line of hide tents and upturned canoes formed the temporary homes of canoemen and voyageurs from Montreal. These were not the reserved Orkneymen of the HBC but brightly dressed French Canadians in blue and red. They sat drinking or singing or arguing loudly by the cooking fires. The less animated, probably recovering from the extended paddle along Lake Superior, lay about the camp.

David made his way through the crowd. Inside, he saw hundreds of men, women, and children. Some were coming to or leaving the various storehouses, lodgings, and huts that occupied the interior. Others were busy unpacking furs or loading barrels onto handcarts. The largest building was the *Cantine* where boisterous *hommes du nord* were drinking rum and gambling loudly. Next to the *Cantine* was the jail mockingly called the *pot au beurre* or buttertub, where those inclined to fight or those too drunk to walk could sleep it off.

David began to have grave doubts about his decision. What could such an unruly company want from him? Were they serious about mapping explorations? Alexander Fraser had greeted him warmly enough at Fraser House last June, he remembered. David was thankful for it too. Leaving the HBC had been a difficult choice, but Fraser had convinced him that the NWC wanted urgently to explore and map the west. "Go right t' the top o' the company, Laddie, and see McGillivray himself," he had told him.

David found the paymaster's office and asked how to find William McGillivray, the senior company partner.

"He'll be in the main house," said the paymaster pointing off to his right. " Can't miss it sir, right in the middle. It's the big log building with the balcony."

Thompson came upon the grand log structure in the centre of the courtyard. Inside, he found a great hall adorned with portraits of the company partners in regal poses and fine garments. Doors off the great hall led to the private apartments of the company's partners. A manservant dressed in a black waistcoat and white collar greeted him.

"I'm David Thompson to see Mr. McGillivray please," said David.

The elderly servant nodded, then, in his most sophisticated manner, ambled into one of the many rooms adjoining the great hall. The servant returned with a slow and deliberate saunter and, mustering up all the solemnity of a courtly occasion, said, "His Honour will see you straight away."

David passed through an open door of varnished oak and entered McGillivray's finely appointed office. Two men rose slowly from a table at which they had been meeting and faced Thompson.

"I'm William McGillivray," said the middle-aged gentleman, the embossed silver buttons glinting on his red tunic. "And this is Alexander Mackenzie," he said, introducing his partner.

David felt somewhat out of place. He had long since given up his European clothes for leather breeches and moccasins, and he had never acquired the mannerisms of a gentleman. The men shook hands and McGillivray motioned for them to sit at the table. The manservant brought a silver tray with three stemmed crystal glasses, each one a third filled with brandy. Thompson declined his drink. This confirmed the rumours that Thompson was a teetotaller. McGillivray was prepared to overlook the shortcoming, although he would find it difficult to trust someone who wouldn't raise a glass with him. *Still*, he thought, *if Thompson is even half as capable as he was reported to be, it will be difficult not to take him on. God knows we need a map-maker.*

McGillivray was nephew to Simon McTavish, the company founder, but he was not put in charge of operations due to his uncle's favour alone. He had worked his way through the ranks using shrewdness and intelligence. He was the company's first non-

French apprentice to winter inland, and he knew the trade intimately. Now he wanted to know more about David Thompson. Could Thompson be relied upon if things didn't go his way? What was driving him to leave a promising career with the HBC? McGillivray knew it wasn't the liberal, if not sometimes debased, lifestyle permitted within the NWC ranks. He doubted it was the extravagance the company afforded its English-speaking supervisors. He couldn't see this thickset and trail-hardened man wanting personal servants to carry luxuries like soft bedding, tableware, and wine along the fur trail. None of that would appeal to Thompson's seemingly Spartan nature. McGillivray even considered that Thompson might still be in the employ of the HBC as an agent sent to spy. The idea was farfetched but could not be entirely ruled out. The shrewd partner would wait and form his opinion after measuring up this quiet man during the course of their meeting.

"Would you prefer wine instead of brandy, David?" asked McGillivray, though he knew the answer.

"No thank you, sir. I don't drink," smiled Thompson.

"I hope your judgment of those of us who enjoy the indulgence is not too severe David?"

"Not at all, Mr. McGillivray, It's just that for me the penalty seems to outweigh the pleasure."

"Then you don't object to the use of alcohol in trade with the Indians?"

"I won't use it for trade myself, but if the NWC stops trading liquor to the Indians, the HBC traders will quickly fill the demand, and if both companies stop selling, then the independent traders and the

Americans will supply even worse grades of cheap liquor and reap the profits. Although I might wish it otherwise, liquor is a principal currency," David answered. He knew alcohol was the leading currency of trade for the NWC.

"The Cree tell me you spend most of the night looking at stars," interjected Mackenzie with a furtive look. "Some say you are possessed by a spirit-man who makes you search the night skies endlessly for some unknown thing."

"They may have a point, Mr. Mackenzie. It takes me many hours to be satisfied with my estimation of coordinates. I'm hoping I can find a sextant with a larger radius to help speed things up."

"And we shall all be the beneficiaries of your efforts, Mr. Thompson," offered Mackenzie. "Your reputation as one skilled in the mathematics and astronomy of map-making is second to none."

"Thank you sir," David said gratefully. The compliment meant a great deal to him. Thompson had yearned for even a fraction of the opportunity the NWC had given Mackenzie. While David plodded along well-used HBC trade routes in the "Muskrat Country," he had heard reports of Mackenzie's exploration to the Arctic Ocean and of his discovery of an overland route to the Pacific. David had longed to be with the great explorer.

"There is little more important to me than making a map of this country," David said earnestly. "Especially of the uncharted lands to the west."

McGillivray had heard enough to satisfy his doubts. "We're in a bind, David," started the senior partner in a sober tone. "We thought that after the Revolutionary War with the Americans, trade would improve. We've increased operations by cooperating with independent traders in the south. The "South Men," as we call them, operate independent of both the American and British influence in the unclaimed territory. Over the past few years we've worked hard to get most of the South Men to join our company rather than competing with us, but now the troubles have gotten worse. The Americans are pushing to claim the area and seem to feel that any part of western North America is their territory."

McGillivray went to a desk and retrieved a bottle of brandy and some papers. He filled Mackenzie's glass then poured himself another as he sat back at the table, papers in hand.

"We've had this dispatch," he said, leafing through the document. "It's from London and warns of discussions between our English masters and the United States. The discussions could result in a border agreement that gives the Americans much of the region. Our London contacts are concerned that the border will follow the forty-ninth parallel to the headwaters of the Mississippi. I wish the damned bureaucrats had consulted us first," McGillivray continued, as his face darkened with anger. "If this damnable agreement is ever ratified, we might have to move all our trading posts in the Missouri country, the upper Red River, and the Mississippi, north of a new border line. We'll be giving up far more than we've gained.

"Trouble is we don't know exactly where the border will be, so we don't know how many trading posts might have to be abandoned or moved. The Americans, curse them, will claim that all the posts are in their territory. To complicate matters, the Spanish are also trying to lay claim to the upper Mississippi. They've sent an envoy to make friendly relations with the Mandans.

"All of this," McGillivray took a careful sip of his brandy and flopped the papers on the table as if discarding a poor card from a bad hand, "and now we learn that there is a new treaty – 'the Treaty of Amity and Commerce' they're calling it – which agrees to the removal of British troops from the southern territory. No troops means no protection from American harassment and nothing to prevent attack from Mackinac Indians. All NWC posts in the territory will be under threat once the troops have gone. Whatever damned border agreement is reached, we know it will mean trouble for us. It's critical now that we gather information to be prepared for whatever comes. First, we need to know exactly on a map where each post is. The South Men never mapped their position."

McGillivray paused and looked squarely at his prospective recruit. "Do you think you could make a map of that region that shows the exact location of all the trading posts?"

"I have no doubt," said David confidently.

"Good, then you are hereby appointed as Surveyor and Map-Maker to the North West Company. You can choose your own crew and mind you outfit yourselves for fast travel. There will be no time for anything other than the task at hand. Fur trading is down on the list of priorities," McGillivray offered.

"Mapping and information gathering, that's what we need from you," he continued, knowing these instructions were what Thompson had been waiting to hear. "And when you're in the Missouri country, investigate the habits of the Mandans for yourself," the wily trader added. The liberated sexual practice of the comely Mandan women was legendary among NWC ranks. Thompson would have no trouble finding a crew.

"We'll expect a full report when you get back here," Mackenzie concluded with a wink as the two partners ushered David to the door.

"But I may not be coming back here," said David. "You may have moved by then."

"Moved? What do you mean?" Mackenzie asked.

"Grand Portage is below forty-nine degrees latitude. It could end up in American territory," said David, matter-of-factly. He excused himself from the room and left the stunned partners to gape at each other, open-mouthed.

Word soon spread, and Thompson was quickly able to select a crew from the endless list of volunteers. The first to line up was Hugh McCrachan, a light-hearted Irishman who had been often to the Mandan villages, sometimes residing there for months at a time. Thompson also chose seven trail-hardened and now grinning French Canadians. Finally, he took on René Jussomme, who could speak fluently with the Mandans, as his guide.

Thompson's party hit the trail in early August 1797. At first they followed the well-travelled route up the Pigeon River, across the Lake of the Woods, then down the Winnipeg River to the vast open waters of Lake Winnipeg. He kept them moving westward until Lake Manitoba, where they paddled north then eventually southwest until they reached the Swan River country. Here he traded their paddles and canoes for horses to traverse the grassy plains.

Days of easy riding brought them to the sweet-smelling woods of the Qu'Appelle River Valley. The aspen leaves were beginning to turn yellow as the party rode south along the parklands bordering the Souris River. During these days of easy going there was time for side trips to company posts. Thompson stayed just long enough at each post to fix a position before moving on. He was still north of the forty-ninth parallel, and while plotting the trading posts he was also filling in valuable gaps in the map of the western region.

November found the survey party at McDonnell's House, about three hundred kilometres from the Upper Missouri and the Mandan villages. Winter had set in and locked the Prairies into a treacherous freeze. It was one of those times when Aboriginal People didn't travel, and wildlife kept still to conserve warmth. Yet they had stayed only a few days when Thompson began packing his gear and loading it onto a dogsled.

"You're not think'n of heading off alone in this weather, are you, David?" McCrachan inquired lazily from the comfort of his fireside chair.

"No," said Thompson, "you're coming with me."

"Has the devil possessed you, man?" protested McCrachan, rising suddenly to his feet. "It's cold enough to freeze the balls off a brass monkey out there. Have you looked at the thermometer? It's twenty below zero!"

"Tell the others we're leaving in the morning. We can make it to the Mandans in about ten days if the weather improves," said Thompson, as he opened the cabin door and walked out to a blizzard of horizontal snow.

The weather didn't improve. The party struggled for days against a stiff westerly that chilled them beyond the already frigid minus fifteen degrees Celsius. The men wanted to stop, but Thompson pushed them farther into the snowstorm. When December came they were still on the trail, and the temperature had fallen below minus thirty degrees Celsius. The men and dogsleds were forced into single file in the unrelenting gale. They needed to keep sight of each other for fear of wandering off track and being lost in the "white death." Often unable to read his compass in the whiteout, Thompson relied on just the wind to tell direction. Only when the branches of some trees brushed his face did he order a halt.

When they finally huddled in a circle in the sparse shelter of a few trees, they discovered one man was missing. Thompson ordered the men back to the incoming trail. They formed a straight line that stretched out into the blinding snow, each man within shouting range of the other. The most distant searcher heard the missing man's faint reply. They lifted his limp body from the frozen ground and helped him back to the protection of the trees and the warmth of a fire.

December ended much the way it had started, with the men trudging across the ice-hardened prairie. Christmas Eve brought clear skies, but they celebrated the day hunkered in a snowbank to hide from a mounted Sioux war party. Thompson had been warned that the Sioux would be searching for plunder or hostages from any travellers they might find unguarded and in the open. Finally, on December 30, Thompson and his men reached the safety of the Mandan villages on the banks of the Missouri.

The Mandans were a peaceable and sophisticated people. They lived in more or less permanent settlements of earthen housing strung along the Upper Missouri. They practised agriculture but thrived mostly as astute traders. They had developed a trading network that had reached far across the continent. It was this infrastructure of tribal trade routes along the continental waterways that the fur traders had unknowingly taken advantage of to develop their own trading empires. The Mandans held annual trade fairs that strengthened strategic trade allegiances with the Crow, Shoshoni, and Cree. They saw the white traders as a new market rather than a threat. Most resourcefully, they enticed NWC traders and native travellers alike by willingly sharing their women as courtesans who would sweeten a weary trader's stay.

Only days after their arrival, Thompson's men readied themselves for the evening dancing festival in the main lodge. They considered bathing, but there was only enough thawed water to wash their hands and faces. Some used a comb to straighten their matted and tangled hair. In the main lodge they arranged themselves amongst the tribesmen in a circle around a large central fire. Chanting and drumming rituals began and lasted for what seemed endless hours to the men impatiently waiting for the Mandan women.

Somewhere around midnight the dancers finally appeared. Stepping in slow rhythmic movements around the fire they smiled and teased the seated men. With time, each dancer chose a man to spend the night with. The choosing was neither random nor haphazard. Some had been instructed to satisfy favoured guests. Others were free to choose. One might select a sleepy old man as a gesture of loyalty to a young husband; another might take two young men with her to encourage jealousy in a suitor. McCrachan and the French Canadians weren't concerned with the reasons for their being chosen, only that they not be left sitting too long.

In the days following the festival, the company of women came at the price of trade goods. When Thompson and his men finally returned to McDonnell's House in January, Thompson's sled was topped with prime fur. McCrachan and the other men had sleds that were near empty.

Thompson spent the late winter on map-making forays along the Assiniboine and Red Rivers. He located several trading posts, including Pembina House and Cadottes House, that were below the forty-ninth parallel and advised them they might be moving north. The slush and sleet of spring slowed him little. He pushed through to the headwaters of the Mississippi and discovered it not to be on the border as believed but rather a full 240 kilometres south. Then, instead of returning on the usual route to Grand Portage, he headed for Lake Superior to survey unmapped territory along its shore. He mapped the lakeshore for 1100 kilometres until May, when he reached a well-provisioned NWC camp.

"David!" a familiar voice shouted as Thompson strode toward the camp's luxurious canvas tent. "What the devil are you doing here?" William McGillivray asked. "Alex, come look who's here," he added excitedly.

Mackenzie emerged from the tent. This time neither Mackenzie nor McGillivray were dressed in their usual finery but were wearing the plain wool garments of the trail.

"What happened?" McGillivray asked, wondering why Thompson had returned to Lake Superior and why he had been away for only ten months.

Thompson, unsure of the meaning of McGillivray's question said, "I thought a shoreline map of Lake Superior was more important than returning down the Pigeon River again."

"Returning from where?" Mackenzie asked.

"The Mississippi," said Thompson.

"You got that far?" Mackenzie asked in disbelief.

"Yes," said Thompson. "We located a dozen trading posts. Some in the Missouri country north of the Mandans, others on the Mississippi. Most have been notified about the possibility of a border, and word will soon travel to the few Houses I didn't reach."

Mackenzie shook his head in astonishment. "You have done more in ten months," he said, "than the best among us could have done in two years."

By the time Thompson arrived back at Grand Portage he had surveyed nearly 6500 kilometres, an accomplishment unparalleled in the annals of North American exploration.

Library and Archives Canada /C-002773/Frances Anne Hopkins.

Voyageurs at Dawn.

The real musclemen. French-Canadian voyageurs paddled furs and trade goods across the continent. With canoes as their only home, they travelled at a pace that could kill ordinary men.

8

Rocky Mountain House

"A hundred pounds sterling for one sea otter pelt!" William McGillivray found it hard to believe. He was presiding over the annual meeting. Seated around a large fir table were the senior partners, ten in all, wearing waistcoats and white leggings and filling the boardroom with pipe smoke.

"Yes sir," assured the junior partner, continuing his presentation to the company's newly appointed chief superintendent and the NWC board.

"Since Captain Cook's report to the Admiralty," the young partner continued, "American and British ships have been trading knives, axes, copper sheets – anything metal – for sea otter pelts with the Indians near Nootka in the Pacific Northwest."

"Yes yes, we knew about that trade," interrupted McGillivray, "but I had no idea prices had gone mad."

"Desire for North American fur has been building in the Orient, sir," responded the young man. "Ships with only a few pelts will sail west and sell them in Canton before shipping back full, laden with tea and spice. The Chinese pay a King's ransom for pelts, and the sea otter is plentiful on the coast. If we can cross overland to the coast, we could sell not only sea otter, but all our other furs from the West, to China at a far better price than our European buyers will pay. We know Mackenzie's route to the coast is not practical for trade, but if there's another river, one that can be navigated with freight canoes, then..."

"Mackenzie!" interrupted one of the grey-bearded partners, removing a long-stemmed clay pipe from his mouth. "He's probably plotting his own plans for the Orient. He was a damned good partner but let's hope he's a piss-poor competitor."

"You're wrong," said McGillivray unable to conceal the anger in his voice. "Mackenzie and company are not our concern. They are operating mainly in the Athabasca region."

There was a hush in the room. Mackenzie's XY Company and Simon McTavish, the founder of the NWC, had been bitter rivals since Mackenzie had broken ranks and gone over to the rival company in 1799. They had been at odds for years before Mackenzie left, and McGillivray had been unable to persuade his old friend Alex to stay. The rivalry was a sore point with the superintendent, and the years of wearing competition were beginning to tell on him.

"But," said McGillivray, regaining his thoughts, "the Hudson's Bay Company has made great gains in the West while we were fighting with XY over muskrats. The Hudson's Bay Company is now in a very strong position."

"It's worth any effort to beat the Hudson's Bay Company in the West. And don't forget the Yanks. If they get a foothold in the West it would take a war to move them out," barked one of the partners.

"I take it we're all in agreement," said McGillivray, bringing the meeting back to order. "I'll send Thompson west to Rocky Mountain House. It's our western trading centre now, and he can mount his exploration from there. If there's no objection, I'll call him in from the Athabasca country and send him out west this summer."

A round of agreement came from the table before McGillivray dismissed the partners. The boardroom emptied. McGillivray's brother, Duncan, remained seated and waited for everyone to leave. Duncan was ambitious and wanted to gain control over the new explorations. One can imagine how the conversation went.

"I want to lead this one, Will," said Duncan to his older brother.

"Thompson will do that well enough for us, Duncan. There's no better pathfinder and surveyor than he," replied the chief superintendent.

"Thompson might be a fine surveyor but he's no clan," retorted Duncan. "Should not the exploration for a trade route to the Pacific be headed by a McGillivray? Is it appropriate that we give David Thompson, a man of no connections, in fact, of no family at all, the honour of finding this vital route?"

"Listen, Dunc," snapped William, "Thompson's one of the finest explorers this country has ever seen. We're lucky to have him, and we know if anyone can find a trade route to the Pacific, it's him."

"Maybe so," said Duncan, backing off a little, "but let's put things in perspective. What he really does is sign up Indian guides to show him the way, and then, with a bit of mathematics and a sextant, he draws a map. I can take on an Indian guide as well as he can, and I'll let him play with his sextant and maps to his heart's content."

"It's no' that simple," brooded William. "Thompson can speak three Indian languages and he knows their ways intimately. As far as connections go, he has them where it counts, with the Blackfoot and Peigan in the West. They know him and trust him. And a bit of mathematics you say! None of us could even come close to learning what he knows. In fact you may not find anyone on this continent who is a better astronomer or map-maker."

"Fine!" retorted Duncan. "I'll keep him at my side when I'm dealing with the Indians and I'll polish his compass for him, but don't sell your own kin short. Give me a chance."

"All right Duncan, you can lead it," said the superintendent reluctantly, "but just how do you plan to explain this to Thompson. He knows he's the best man for the job."

"Why not promise to make him a partner, if you think so highly of him, or tell him we'll give him time to work on his cherished map when he returns."

"All right, but mind you listen to what he says and follow his advice or..."

"Agreed, agreed," interrupted Duncan with a wave of his arm. Then he leaned back in his chair, satisfied.

Duncan McGillivray set out for Rocky Mountain House leaving Thompson to follow later in the season. The inexperienced explorer doggedly tried for more than a year to find a route through the impregnable mountains. Every promising defile and river valley he chose to follow led to treacherous slopes and jagged crags passable only to mountain goats and eagles.

When Thompson arrived, McGillivray was still in the mountains, entangled in his desperate struggle to be first to find the pass. Thompson, trail-wise and wanting to avoid the senseless competition with McGillivray, tried a different approach. He had heard from the Peigan that a band of Kootenay lived on the headwaters of a great river that flowed to the sea. If he could find some Kootenay on this side of the mountains, they might point the way to their distant tribesmen, and Thompson would find the pass and his river. He mounted a solid horse and rode into the Bow River country on the lookout for Kootenay tribesmen. He visited many Peigan camps and shared their fires in hopes of finding his guides. But the Peigan were tight-lipped. The Kootenay were their old enemies. While open warfare had ceased and there was now some trade between them, the Peigan kept the Kootenay at the margins of Peigan territory. Trade occurred only when the Peigan wanted horses or furs from the Kootenay to supply the NWC. Any direct trade across the mountains would jeopardize the Peigan stronghold, and worse, put arms into the camp of their old enemies.

"Would you place guns in the hands of those who would kill us?" a young Peigan chief admonished Thompson as they smoked around a fire. Thompson knew better.

"The Kootenay already have guns, and they came from you in trade, so you cannot ask me to stop what your people have already done," he responded truthfully.

When Thompson finally located a small group of Kootenay, they were near starvation. They had been under constant harassment from Peigan warriors and only the local Peigan chief had prevented them from being killed outright. Thompson intervened and was able to supply the Kootenay with fresh horses and provisions. He would accompany them across the mountains to the river tribe. But the horses and provisions he supplied were soon stolen, and no matter how often Thompson tried to re-supply the Kootenay under his protection, the Peigan would just as quickly steal the supplies away. Eventually the Kootenay quietly slipped off in the night, leaving the troublesome Peigan behind for good. Thompson stayed behind as a diversion but sent two voyageurs with the Kootenay in hopes of learning their route. The Kootenay and the two voyageurs were never heard from again.

Duncan McGillivray was probably not disappointed. His own search was over. His frantic pace of exploration and his desire to find the route by himself had ended with a severe attack of rheumatism and physical exhaustion. McGillivray had little choice but to return defeated to Fort William, which now replaced Grand Portage as headquarters. Still unwilling to give Thompson the lead, he put James Hughes in charge of any remaining explorations. He also supplied Hughes with a Cree guide known by both McGillivray and Thompson to be unreliable. The results were not unexpected, and in 1802, with no route found and with reports provided by McGillivray, headquarters called off exploration for the Columbia. Thompson was ordered back.

ꝏ

At Headquarters Thompson was told Mackenzie's bustling enterprise, the XY Company, had been giving serious competition to the Northwesters, and they needed him back in the Muskrat Country to bolster trade. He took the setback in stride. There was little to interest him in that swamp-strewn landscape, but he was rid of Duncan, and there was Charlotte now.

He had met her at Isle-a-la-Crosse, an NWC factory run by Patrick Small. Small, an influential wintering partner, had recently retired to England, and as most did, he left his native wife and their children behind. Small abandoned two daughters, Nancy and Charlotte. David had passed a comfortable night in the factory's bunkhouse, but he had slept an hour late. The soft bed was a luxury after days spent sleeping on the cold ground under his canoe.

He first saw Charlotte Small when he was hurrying from the bunkhouse to load his canoe. It was her back he noticed. She stood at a table with two other girls scraping reeds. He had never seen a back as lovely as that one. It tapered gently from her soft round shoulders down to her perfect waist. She turned her head over her shoulder and smiled quizzically at the intense young man staring at her.

Charlotte's inquiring smile, her long hair, her dark eyes – everything about her struck him as surely as if he had just walked into a tree. Charlotte and the girls, their baskets full of cleaned reeds, disappeared into the cookhouse. Even the way she moved, with a feathery lightness, kept David rooted where he stood. Only after she had gone was he able to shove his canoe from the rocky beach and paddle toward his bush camp.

In the days and nights ahead, he could not remove the image of Charlotte from his mind and his obsession with her only grew. To him, she was the coming together of two worlds. Her dark eyes had revealed the stoicism and good nature of her Cree mother but glimmered with the passion of Celtic ancestry. Auburn hair, a blend of native raven and Irish red, lay across her shoulders. He knew her father's famous family had given generals and commanders to the British Empire, and the pride and intelligence from that lineage he saw in Charlotte as quiet dignity.

ꝏ

In the months ahead, David Thompson began to make frequent astronomical observations at Isle-a-la-Crosse. It was for his map, he told himself. He returned again and again to add details and fix and re-fix positions by the stars. The outpost was in danger of becoming the most charted location in British North America until eventually, in June of 1799, he asked Charlotte Small to marry him. The ceremony was simple and in keeping with local customs. He offered gifts to Charlotte's relatives and then declared, in front of those assembled, that she was now his wife. Like many country brides in that era, she was young, only thirteen, but they remained together for over fifty years, to be parted only by death.

In 1804 Thompson was summoned from the Muskrat Country to Fort William, where William McGillivray made him a partner in the NWC. The superintendent knew Thompson was more than deserving of the reward. Few NWC men had contributed as much as David Thompson to the company's fortune. The reward he most wanted was the opportunity to map the West. Unfortunately, the company chose otherwise and sent him back to trading duties in order to ward off the competitive advances of the HBC. This was a critical mistake. The United States government took advantage of the confusion north of the border and dispatched Lewis and Clark across the continent to the Pacific. Armed with full government support and Thompson's maps of the Missouri country, the expedition found the lower Columbia and followed it to the sea in 1805.

The NWC again shifted its focus to the West. The lower Columbia was navigable and they wanted access to the river from the north. The Americans' next move in their quest for control of the Northwest would be to establish forts along the river. McGillivray wanted to beat them to it by building NWC forts and by encouraging trade relations with the Aboriginal Peoples before the Americans became established. He recalled Thompson to headquarters and ordered him west. He instructed him to find a usable trade route through the Rockies, then build trading posts on or near the Columbia. This time, Thompson would be in charge of finding the route and building the posts.

Thompson left the Muskrat Country for the final time in June 1806 and once again travelled across the broad expanses of prairie grass to the shadow of the Rocky Mountains. He arrived, with Charlotte, in the spring and reported to John McDonald of Garth.

"Davie!" bellowed John as Thompson rode into McDonald's camp at Rocky Mountain House. "Good ta see ye man, and Charlotte too." McDonald was in command of the Northwesters' Prairie operations. He respected Thompson and knew him as well as anyone did. They were, in fact, brothers-in-law. McDonald had married Nancy Small. "Nancy, come, yer sister is here. And what's this, Davie?" said John, looking at Charlotte, whose belly was conspicuously swollen with child. "So it's not true after all. You dinnae spend all yer nights just gazing at the sky."

That evening McDonald told him of the increasing troubles with the Peigan. Some traders had been killed and both the Peigan and Blackfoot were suspected. The tribes knew of the NWC plans to expand over the mountains and had escalated their campaign of deterrence with increased harassment and threats. Thompson had learned enough from past failures and his talks with the Peigan to know that to try to cross the mountains directly now would spell disaster. He had also learned enough to narrow down the options for locating a mountain pass. He suspected it was along a route some Kootenay had used to return home years earlier. Under close scrutiny from both the Peigan and HBC traders, he decided to send his trusted mixed-blood packer, Jaco Findlay, alone to reconnoitre the possible route.

Findlay returned in the spring. As predicted, he had found the pass, complete with evidence of old Kootenay campsites on the far side of the high mountains. Thompson waited, thinking of ways to cross without detection, until the American officer Captain Lewis unknowingly provided the opportunity. Lewis, of the Lewis and Clark expedition, had provoked an attack by the Blackfoot, in which two warriors were killed. The incident enflamed the Plains tribes, and the Peigan joined Blackfoot war parties heading south to avenge the deaths.

Thompson, finally free from prying eyes, took the opportunity to move out. He packed up his family, Jaco, and the other men, along with all the supplies and equipment needed for trade and advanced toward the forbidding mountain peaks. Thompson led the climb over the foothills and into the treeless barren rock for which the Rocky Mountains were named. His party struggled into the rarefied altitudes, clinging to stunted shrubs, stumbling and backsliding across the loose scree and talus slopes. Eventually, horses and men, women and children descended into a high valley through which streams trickled west to the Pacific. They followed the rivulets to a larger stream and finally to a river in a wide valley. David took them upstream, and in the weeks ahead they built a cabin at a site suitable for trade. They didn't know it was the first trading post built on the Columbia.

Thompson sent a small party back to retrace their route and bring back more supplies for the coming winter. The party returned in early fall with troubling news. Old White Swan and his Blackfoot had sacked the NWC trading post at Fort Augustus. They had captured a large cache of arms and ammunition along with all their other plunder. Tribal aggravation over white trade with the Kootenay had reached a murderous pitch. They had also learned that Thompson had crossed the mountains. He ordered a stockade built to fortify their position, but two Peigan scouts arrived before it was even completed. Fortunately, the advance party had come on foot, and they were not well armed.

Thompson's men worked nervously to finish the stockade, which now protected the post on three sides. The back side was protected by a bank that descended steeply to the river. At night they could haul water up the bank unmolested by using a bucket and rope. Those on guard duty shoved loaded guns through holes cut into the fort walls. By the time the main Peigan war party arrived, the post was well fortified and provisioned with dried meat and plenty of water. There were six men with ten guns between them inside the fortifications and forty Peigan outside its walls.

Thompson knew that the Peigan had held a council and were determined to destroy the fort and everyone in it. He remembered his first years on the plains twenty years ago with James Gady. He remembered his time with old Saukamappee and how he had come under the protection of the great Peigan war chief Kootenae Appee. David Thompson probably knew these Plains people, and they knew him, better than any white man alive. Many of the warriors were children when Thompson first came west to the Peigan camps. The fact that they had not already attacked attested to his relationship with them.

"Tell your people," he told the Peigan leader, "that many of you will die fighting against wooden walls. Is it not better to have these?" he asked, handing the chief several large tobacco strings and many pipes with sought-after long, ornamented stems. "Give these to your chiefs and tell them, even if you kill us, we are strong under King George and will always have more guns."

The Peigan returned with the offerings to a high council where some of Thompson's gifts found their way into the hands of the elderly war chief, Kootenae Appee. He had been trying to dissuade his tribesmen from killing the white men, but Sakatow, the civil chief, wanted blood. When Kootenae Appee received the gifts from his old friend Thompson, he smiled. He now had what he needed to persuade the others. A large war party of about two hundred had crossed the mountains and was poised to strike from their camp only thirty-two kilometres from Thompson's fort. Kootenae Appee assembled the chiefs, and Thompson's gifts were distributed.

"What can be done with these?" he asked, referring to the generous supply of tobacco and pipes. "If we attack, nothing of what has been given us can be accepted."

The oldest of the three war chiefs eyed the tobacco wistfully, for he had gone without any for many months.

"I have attacked tents," he began, "my knife has cut through them, and our enemies have been killed and I am ready to do so again. But to fight against logs that a musket ball cannot go through, and against people we cannot see but whom we know. This is what I am against. I go no further."

The old chief cut a plug of tobacco and filled a pipe. He handed it to Kootenae Appee, saying, "It was not you that brought us here, but the foolish Sakatow who himself never goes to war."

They all smoked, took the gifts, and then returned to the plains before the snows of winter blocked their passage over the high mountain passes. It was not the last time Kootenae Appee would keep his vow to protect his old friend.

David Thompson returned to the business for which he had been sent. He set up a brisk trade with the Kootenay at the fort, which he named Kootenay House. The following year he travelled south by canoe and horse into what is now Montana and Idaho. Within six weeks he completed a remarkable exploration of nearly a thousand kilometres before returning to Kootenay House for the winter of 1808. He was a partner now and would share in the profits of his trade. He wanted to secure enough money to support his growing family. He knew his trading days would soon end. It was a physically demanding life, and no one grew old on the trail. They either died or retired early.

When the snow cleared, he once more travelled south. This time he built two trading posts, Kullyspell House and Salish House, in the Columbia basin. After securing trade arrangements over the winter with the Salish and Flatheads, he packed his family and a large load of fur onto the horses and headed back east across the mountains to Rocky Mountain House. He had found a pass through the mountains, built three new trading posts in the fur-rich West, and established trade with the western tribes. Although he didn't know it, he had also built the first three trading posts on the Columbia River and its tributaries.

Painting by Joe Cross/Courtesy Wildsight, http://www.wildsight.ca

Thompson Crossing the Athabasca Pass, January 1811.
With eight dogsleds and thirteen men, only half his original complement,
Thompson pressed on to do what some thought impossible.
He crossed the high Rockies in the dead of winter.

9

The Columbia

On a cloudless spring day in 1809, David Thompson's head fur trader, Finan McDonald, steadied the barrel of his long gun on the top ledge of the makeshift barricade. Beside him were the voyageurs Michel Bourdeux and Alexis Bouche. They joined 130 Salish warriors with bows drawn and guns ready. All waited silently behind the hastily built fortifications for the Peigan to attack. Behind them, the women and children lay hidden on the ground.

Two days earlier, Thompson and the Salish had entered "the war grounds." This region had been the traditional buffalo-hunting place of the Salish until the Peigan had driven them out. But now the Salish had NWC guns and had returned to hunt buffalo for themselves and for Koo Koo Sint. This was the name the Salish had given Thompson. It meant the Man Who Looks at Stars.

Peigan scouts discovered the hunting camp and returned with many warriors on horseback to drive back the intruders. As past battles had shown, it was not the number of warriors but the number of guns that determined the outcome. On this score, unknown to the Peigan, the Salish were equal. Not only were they armed, but they had also adopted the traders' way of shooting. They knew to take precise aim while supporting the musket across their knees or over a log. The Peigan, on the other hand, made war as if the Salish were buffalo being hunted on the plains. They galloped forward, shouting wildly and firing from their charging mounts. This time, however, their fearsome war cries were silenced, one by one, as the Peigan and their horses fell in the deadly hail of carefully aimed musket fire.

When the battle was over, the Salish had regained their old hunting grounds. The Peigan fears of a well-armed enemy had been realized, and David Thompson was no longer safe. Word spread among the Plains tribes, and Koo Koo Sint and the NWC traders were now marked for death.

Thompson and his men skirted back into the safety of Salish and Kootenay territory, but they had to make their way over the Rockies to return home. The Peigan knew the passes well, and Thompson's brigade, now swollen to sixteen horses loaded with forty-six packs of fur, would be lucky to evade detection. On the morning of June 9, 1809, Thompson brought the horses to an abrupt halt. They had reached McGillivray's Carrying Place, the first staging area for the ascent over the mountains. It was a place they had used on previous crossings, but someone had already been there and left fresh tracks. On the ground lay a broken aspen branch with the bark peeled back. Thompson knew this to be the sign left by Peigan war parties proclaiming their presence in enemy territory. The tracks, only hours old, were headed for the mountain pass.

The brigade unloaded the horses. The animals could not carry their heavy burden over the perilously narrow mountain trail. Much of the gear would be shouldered by the men until safer footing for the horses could be reached. Reading the tracks, they stayed several days behind the Peigan until they descended to the headwaters of the Saskatchewan River. Here they reclaimed stashed canoes and repaired the bark seams with pitch and tree gum before heading downstream. For safety from attack, they camped on the narrow ledges above the river, leaving the canoes below.

After two days they passed the ruins of Fort Augustus. The fort's remains served as a further reminder of the Peigans' anger. It was June 22, thirteen days since they left McGillivray's Carrying Place, and they had been fortunate to avoid Peigan scouts. With each passing day the threat of attack diminished, until exactly one month later they arrived, exhausted but safe, at the old trading post at Rainy River.

Thompson, now reunited with Charlotte and the children, was more than ready for his rotation down to Montreal. Wintering traders were given time to return east and recuperate from the hardships of the trail. Thompson's time was two years overdue. He was exhausted and looked forward to the quiet intimacy Charlotte and his family provided. Unlike anything he had known previously, this intimacy soothed the unrelenting loneliness that had been with him since his days at Grey Coat. Thompson and his family climbed into light canoes and headed east on the Rainy River. They had paddled for only a few hours when they were intercepted by NWC couriers with a dispatch that contained orders for Thompson to return to the Columbia. This time he was directed to follow the river to its end.

The NWC had been vacillating over its interest in the Columbia and a route to the Pacific. Exploration was expensive. Mackenzie had failed to find a usable route, and Simon Fraser's recently travelled path down the river that bears his name was equally untenable for trade. The company was ready, once again, to abandon the search. Then they heard that the American, Jacob Astor, and his newly founded Pacific Fur Company were planning to build a Pacific trading port at the mouth of the Columbia.

The Northwesters and Astor had been negotiating to share the profits from the Columbia region through a trade agreement. The negotiations were a ruse, only used as a stalling tactic by both sides to provide sufficient time to organize their own competing expeditions. Astor, with the backing of the United States government, had the advantage. William McGillivray, on the other hand, had been unsuccessful in getting British Foreign Office support for either overland or seaborne endeavours. The British, despite pleas from the Northwesters, would not consent to dispatching even one of His Majesty's ships to challenge the Americans at the river's mouth. The NWC would have to go overland alone, and for this they would need David Thompson.

Thompson was too aware of the dangers ahead to give in to Charlotte's tearful demands to accompany him. Their canoes went separate ways, and Thompson returned to the post. He was supplied with men and four canoes loaded with trade goods. He was to paddle to the headwaters of the Saskatchewan, trade some canoes for horses, and once again cross the mountains by the only route known to them.

It was a route also known to the Peigan. Forty warriors had pitched their tents on the north bank of the river waiting to intercept any traders going upstream. Thompson knew he would be hunted and his chances of slipping through the Peigan blockade would be slim, but he pressed on. He was already late if he wanted to reach the Pacific that season.

Light provisions required for a mountain crossing meant that they would have to risk hunting to supplement their food rations. Thompson, once he had horses, went ahead with a mounted hunting party. They left the main brigade and canoes behind with plans to rendezvous farther up river. Game was plentiful in the mixed forest and meadows of the foothills and, in a short time, they killed three elk. Toward evening they had nearly finished caching the meat on a raised platform when one of the older Cree hunters turned to Thompson. The hunter was ashen with fear.

"I have had bad dreams," he said, pointing to the cache of meat. "That meat will never be eaten; I must leave you."

The old Cree then collected his horse and rode away into the fading daylight. Thompson knew not to dismiss the warning as idle superstition. Native dream stories, if not premonitions, were at least an indirect way of disclosing what they knew was coming. When the brigade failed to arrive the next day at the appointed time, Thompson sent William Henry and a young Cree downriver to find them.

"Go quietly and keep out of sight," Thompson cautioned. "Don't fire any shots unless it's to defend yourselves. The Peigan must be close."

About twenty-four kilometres downstream, Henry discovered the Peigan blockade perched on the bank to guard the river below. He skirted past their tents and returned to a safe distance farther down the river. Here he found evidence of the brigade from tracks and canoe keel marks left in the sand. He also found a quickly built stone barricade and traces of blood splattered across the stones. Henry assumed there had been some kind of standoff and the voyageurs had escaped downstream. He fired a signal shot in hopes of alerting the voyageurs to his presence but there was no answer. He followed the river for some distance but not finding the canoes he returned to Thompson that night and recounted his findings.

Thompson was fuming. "You fired a signal shot! Why didn't you just ride into the Peigan camp and announce our presence?" he asked with blistering sarcasm. "We'll need to move out before first light. Chances are they followed you."

Thompson and his men set off in the predawn blackness for the Brazeau River, a north branch of the Saskatchewan. By daybreak, the Peigan had arrived at his abandoned camp. They followed Thompson's tracks into the forested hills and through a maze of fallen trees. The distance between them was closing until, in late afternoon, the dark grey skies filled with white flakes of an early snowfall. The skiff of accumulating snow was not enough to conceal their tracks until three grizzlies intervened. The bears, attracted to the scent of Thompson's trail, sniffed and pawed the ground, eventually sitting on his tracks.

The Peigan, intent on the fading horse trail, approached too close to the bears. The largest grizzly reared up and bared her formidable teeth. The Peigan retreated but the bears did not. They sat on the trail long enough to convince the warriors that Koo Koo Sint had sent the bears to guard his path. By evening, the snow had obscured all traces of Thompson's trail, and the escaping party risked a small fire and tried to sleep.

Thompson could not sleep. He knew the morning would bring a renewed chase. He needed to find his lost voyageurs and resume his journey, but he also knew the Peigan would pursue him wherever he went. He remembered the warning of the friendly chief, Black Bear: that "Four war tents have been sent to kill you to avenge the Salish victory." He, David Thompson, was the object of their vengeance. He knew his life was still in grave danger.

Before dawn he had worked out his plan. He sent Henry and the others ahead to find the voyageurs and return them to Rocky Mountain House, where he would meet up with them later. Thompson then went into hiding alone. He found a thickly forested hill and, covering his tracks, he holed up in his carefully camouflaged camp. Henry and the men would have a better chance of surviving the Peigan without him.

Henry found the voyageurs resting at an abandoned trading post down river. He persuaded the anxious men to return with him to Rocky Mountain House. For nearly a month there was no sign of Thompson. Finally a search party discovered him – cold, starving, and delirious – near his hidden camp. He was in the advanced stages of exposure. He refused to recount what happened during his month in hiding, and the men knew better than to press the question.

By late October, 1810, Thompson had recovered and was planning the unthinkable: crossing the Rockies in winter. The Peigan too were planning. They would put an end to all cross-mountain trade. The Plains tribes had assembled in an impressive coalition of Peigan, Blood, and Sarcee. They had encircled Thompson and the NWC post at Rocky Mountain House and were determined to halt any further trade over the mountains. The ostensible purpose of their gathering was trade, but recurring war songs and the intimidation of the traders left little doubt of the real menace they represented. Some chiefs wanted to see the white man's blood spilled in the prairie snow. Others, Kootenae Appee among them, were for negotiation. The aging war chief was protecting his old friend inside the trading post but he also foresaw the danger to his people if they began war on the NWC men.

For Kootenae Appee, however, the most serious threat to his people was not guns for the Salish or loss of trade monopolies, but alcohol. The effect alcohol was having on his people was devastating. It had uprooted moral and cultural norms and reduced native men and women to the worst corruption and degradation. Many of the assembled warriors had come not to support the tribal cause but for the free liquor the traders were using to buy a temporary truce. He had seen all this, and he knew that the only trader who would not poison his people with drink was Koo Koo Sint. Thompson had destroyed kegs of NWC liquor rather than use them for trade. He had not forgotten his childhood experiences, but even the drunken misery of London's Whitechapel or St. Giles seemed restrained in comparison to the debasement he had witnessed among the Aboriginal Peoples.

William Henry, however, had no reservations about alcohol. Hoping to get Thompson out during the drunken confusion, he had twice distributed large kegs of liquor to the tents outside the post, but even that had not been enough. Finally, Henry laced several kegs with the painkiller laudanum, and Thompson and his men slipped through the tents unnoticed. It was futile to attempt a mountain crossing by the usual route up defiles of the Saskatchewan. The brigade travelled north instead of west along the usual Saskatchewan route. This, plus the belief that no man would be foolish enough to try to cross the mountains in the winter, prevented the Peigan from following Thompson's trail.

On October 28, 1810, Thompson led his reluctant brigade toward the Athabasca, the other great river whose headwaters penetrate the western approaches to the Rockies. There were twenty-four men and the same number of fully burdened horses. Many of the men were anxious about the winter crossing, and the hardships of cutting a trail for almost a month through windfalls and thick forest did little to improve their mood. With the horses spent and game scarce, one among them, Du Nord, was for turning back. Thompson remembered the words of Captain Tunstall that day, twenty-five years ago, on the decks of the *Prince Rupert*. "You don't know the measure of a man till seen in hardships." The hardships for himself and his men, Thompson knew, were just beginning. He took temporary reassurance from knowing the stamina of the French speaking voyageurs was legendary.

They finally emerged from the forest and onto easier going across the frozen marshland of the Athabasca plain. But this did little to cheer Du Nord, who was now agitating among the crew for desertion. The men were hungry and the temperature had dropped to minus thirty-four degrees Celsius. As they travelled up the Athabasca toward the mountains, the snow deepened and the horses became bogged down. Some were killed for food and others were traded for dogs. The men built crude dogsleds from logs and continued to climb. Du Nord, overcome with resentment, began clubbing the dogs mercilessly. By the time Thompson angrily stopped him, one dog lay dead. Fortunately, Du Nord and several other troublesome voyageurs deserted before Thompson had to quell the widespread unrest that was surfacing in his crew.

The dogs strained against the sled traces and the men floundered in their snowshoes as the brigade laboured into deepening snow. The heavy loads sank into the steep-sided snowdrifts, and Thompson was forced to send some men and trade goods back. Those who were sent did not complain. He continued into the frozen heights with eight dogsleds and thirteen men, only half of his original complement. The brigade, pitifully weak, trudged toward the ice-clad peaks, offloading some heavy gear as they went in hopes of recovering it later. Here, in mid-January, in one the most treacherous and frozen places on earth, nature's benevolence intervened. A soft Chinook wind, which sometimes sweeps this winter landscape, enveloped them in warm air. The temperature miraculously rose to zero degrees C. When the cloud-enshrouded peaks cleared, the men found themselves at 1500 metres on the summit of Athabasca Pass. Below, in the distant valley, lay the headwaters of the Columbia.

What had taken weeks to climb took only days to descend. Yet, for some, the prospect of continuing to the Pacific was overwhelming. More men deserted. With only a few men and limited supplies, Thompson had no choice but to wait until spring before he pushed onward. Some faithful voyageurs were sent back to the summit to bring forward the offloaded supplies. Some hunted game and laid away deer and moose meat while others built cedar boats for the run down the river. They had found a safe mountain passage, away from the marauding Peigan. The Athabasca Pass was to become the route over which fortunes in fur for the NWC and later the HBC would be hauled.

Thompson no longer had enough men or trade goods to establish a trading post at the mouth of the Columbia. They barely had enough arms and provisions to care for themselves, never mind extras to pacify any hostile native people they might encounter. In April, 1811, they headed for Thompson's old post at Salish House to re-supply before descending the Columbia.

Spring travel brought its own hardships. Wet snow and heavy rains soaked them. Spring runoff dangerously swelled the rivers. Ice threatened to break as they dragged heavy gear across thinly frozen lakes. By May they had acquired horses to lighten their burden, and by the end of that month they reached Salish House. It took another month to re-supply, cross overland, and build a boat. Finally, on July 3, they shoved off and drifted downstream on the lazy currents of the Lower Columbia.

David's thoughts, as if encouraged by the flow of the river, were now flooding with memories. From a boat not too different from this one, he had first stepped ashore in a new land so many years ago. That Grey Coat apprentice was now about to complete a mapping journey that spanned the continent from the banks of the Churchill River on Hudson Bay to the mouth of the Columbia on the Pacific. He recalled the brief kindness Captain Tunstall had shown him as a boy on *Prince Rupert*'s decks. He thought of Hodges and Mr. Prince teaching him how to shoot and hunt at Churchill Factory. He remembered the generosity of

Philip Turnor and the great gift that kind man had given him. Through his travels, he had held the memory of these encouragements, however brief the encounters had been, close to him. He was thankful too, to have lodged with old Saukamappee and for the faithful protection Kootenae Appee had given him. These men had helped him as much as any other had.

As he drifted closer to the river's mouth he realized his days of exploration were nearly finished. On July 15, 1811, he caught scent of the sea, and a few hours later the boundless expanse of the Pacific opened before him. He was not surprised to see log huts on the estuary's shore. Salish tribesmen upriver had told him the Americans had come by ship just months earlier. His task was to set up competing trading posts and map the trading route. This he continued to do, but more than that, he had found the river route that Mackenzie and Fraser had sought in vain. His accomplishments on land now rivalled what Captain Cook had done at sea. When he finally retired east a few months later, in 1812, to finish his map, he had travelled some eighty thousand kilometres on foot, horseback, and canoe. David Thompson had charted a continental area so vast it remains a mapping achievement unparalleled in human history.

Today a monument marks where David Thompson is buried in Mount Royal Cemetery, Montreal, but for many years his grave lay unmarked.

Charlotte Small is buried beside her husband.

Epilogue

Unfinished Work

David Thompson's detailed information on the Canadian landscape was so complete that mapmakers were to use his observations for more than a hundred years, yet none of his achievements bear his name. Only Simon Fraser honoured him by naming a tributary of the Fraser River after his modest friend. The NWC, Alexander Mackenzie's XY Company, and the HBC eventually merged under the Hudson's Bay Company banner. The Hudson's Bay Company saw Thompson as a traitor and had no wish to advance his name. His painstaking cartography was taken by the HBC and incorporated into their maps of British North America. They ignored the man who had produced them.

In his later years government officials mistrusted Thompson and unfairly blamed him for the loss of the Columbia Territory to the United States. Some historians today still blame him. The man who, more than any other, had explored and documented the vast reaches of Western Canada was forgotten and died in poverty.

His journal remained unpublished for sixty years. It was eventually found by Joseph Tyrell, a geologist for the Canadian government. Tyrell, recognizing the priceless value of Thompson's unfinished work, edited the manuscript and had it published in 1916. Many years later, David Thompson's grave was finally marked. "DAVID THOMPSON 1770-1857. To the memory of the greatest of Canadian geographers who for 34 years explored and mapped the main travel routes between the St. Lawrence and the Pacific."

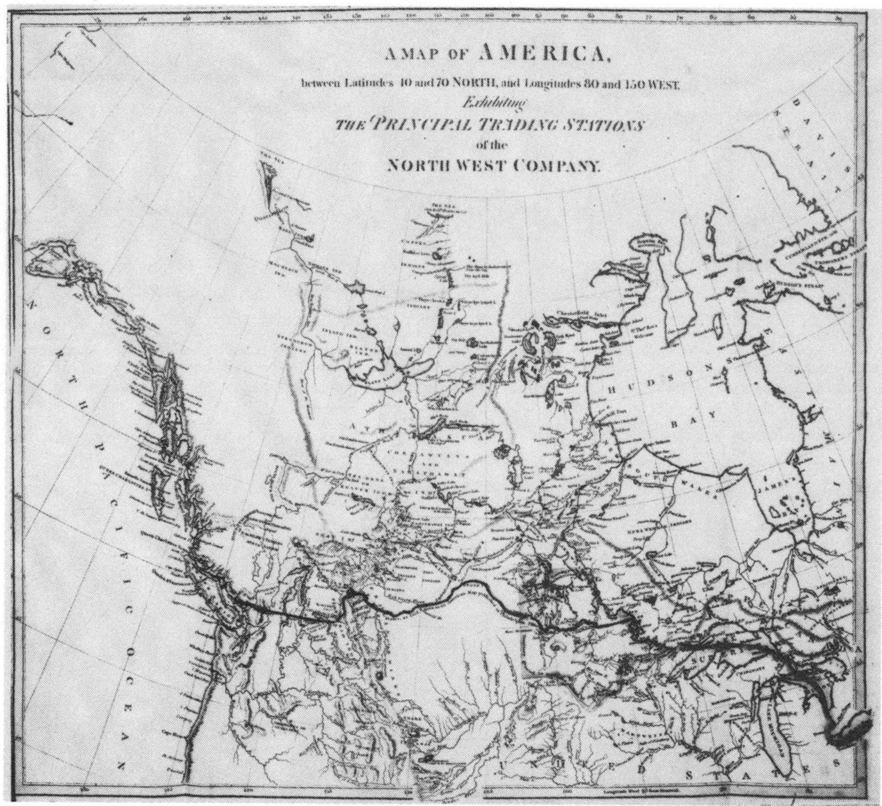

Library & Archives Canada/NMC 6763.

Thompson devoted his life to mapping Western Canada. It was an accomplishment comparable to Captain Cook's famous charting of the Pacific Ocean.

Chronology of David Thompson (1770-1857)

Compiled by Clarence Karr

DAVID THOMPSON AND HIS TIMES	CANADA AND THE WORLD
	1670 King Charles II of Britain grants a royal charter to the Hudson's Bay Company for exclusive trading rights in Rupert's Land, the drainage basin of Hudson Bay. Neither the monarchy nor anyone in the company has any idea that the charter territory would include most of what is today the prairie provinces and northern Ontario.
1770 David Thompson is born in London to Welsh parents of little means.	**1770** Samuel Hearne of the Hudson's Bay Company (HBC) departs on his third expedition of discovery in the arctic region of the North American continent.

DAVID THOMPSON AND HIS TIMES

1772
David Thompson's father dies, leaving his widow and their children in poverty.

CANADA AND THE WORLD

William Clark of the future Lewis and Clark expedition is born.

Captain James Cook claims the east coast of Australia for Britain.

British troops fire on civilians in Boston killing five American colonists.

1771
James Cook completes his first voyage of discovery around the world.

Oxygen is discovered as a chemical element.

1772
Samuel Hearne reaches the Arctic Ocean via the Coppermine River.

1774
The HBC, responding to intense competition from the Montreal-based fur traders, establishes its first inland fur trading post at Cumberland House on the Saskatchewan River.

1776
With their Declaration of Independence the rebel Americans proclaim a new nation based on new liberal principles of equality and freedom. Those remaining loyal to Britain or trapped in a supportive position flee home to Britain, to the British West Indies,

David Thompson and His Times	Canada and the World
	or north to the British colonies of Nova Scotia and Quebec.
1777 David Thompson enters the Grey Coat Charity School, where he will receive good technical training. This will facilitate his career as a surveyor and map-maker. The school prepares some children for careers in the Royal Navy.	**1777** In the American Revolutionary War, French Marquis de Lafayette assists in the training of the American army. The Second edition of the *Encyclopedia Britannica* is published.
	1778 Captain James Cook explores the Northwest Coast of North America in search of the Northwest Passage. Louis XVI of France, in aid of the American revolutionaries, declares war on Britain.
1779 As a child David loves reading adventure stories such as *Robinson Crusoe* and the *Arabian Tales*.	**1779** Mohawk chief Joseph Brant leads a group that destroys the Goshen, New York militia at the Battle of Minisink in the American Revolutionary War. Hawaiian natives murder Captain James Cook.
1783 Young David Thompson has seen the devastating effect of alcohol on London's poor. These experiences influence his eventual decision to live a life without alcohol.	**1783** The Treaty of Paris officially ends the American Revolutionary War and results in the recognition of the United States of America as an independent nation.

David Thompson and His Times	**Canada and the World**
	The North West Company (NWC) is formed by a group of Montreal merchants to manage the fur trade in the British territory to the north and the west.
1784 In June Thompson embarks on the HBC's ship *Prince Rupert* as an apprentice to the fur-trading company for a seven-year period. He will never return to England or see his mother again. Thompson serves under Samuel Hearne at Churchill Factory on Hudson Bay.	**1784** Captain James Cook's journal of his last voyages appears from a London publisher. The Russians establish their settlement in Alaska with a fur-trading post on Kodiak Island. The HBC acquires many of its workers from orphanages in this period, in hope that their young age will result in longer period of service. The American ship the *Empress of China* sails to Canton. This is the beginning of the China trade for the new United States (U.S.) Thousands of refugee Loyalists from the American Revolution result in the division of Nova Scotia into two colonies with the appearance of New Brunswick.
1785 The HBC sends fifteen-year-old Thompson to York Factory, its port of entry on Hudson Bay at the mouth of the Nelson River. Thomson and two Chipewyan guides walk the 240 kilometres (150 miles).	**1785** The U.S. chooses the dollar as it official currency.

David Thompson and His Times

1787
Thompson journeys inland to establish the trading post South Branch House on the South Saskatchewan River and assist other posts.

He learns the Cree language during this period, a necessary skill for any successful trader.

He winters with the Peigan people in the Rocky Mountain foothills on the Bow River.

1788
Thompson seriously fractures his leg in a sled accident at Manchester House. His recovery will be a long one.

1789
Thompson continues his recovery at Cumberland House. Here he meets Philip Turnor, the famed HBC surveyor and prominent man of science. Part of Turnor's duties is to train a select few HBC men as surveyors. Thompson begins his training.

Canada and the World

1787
British admiral and explorer Francis Drake dies.

Americans gather at Philadelphia to write their constitution.

Captain Arthur Philips leaves London with eleven ships to establish a British penal colony in Australia. A new location is required because of the loss of the American colonies.

1788
Captain Arthur Phillips arrives in Australia and builds Sydney as the first settlement on the continent.

1789
NWC explorer Alexander Mackenzie, hunting for an inland sea that would lead to the Pacific and provide a good transportation route, reaches the Arctic Ocean via the river that will bear his name. The inland sea, long sought after, did not exist.

The French Revolution begins.

George Washington is elected first president of the United States.

1790
British Captain George Vancouver begins a three-year survey of the Northwest Coast of North America.

David Thompson and His Times	**Canada and the World**
1791 Thompson completes his seven-year apprenticeship. He accepts the usual three-year HBC appointment. He begins trading and surveying in Muskrat Country, a large swampy area infested with mosquitoes, and one of the regions least favoured by HBC workers.	**1791** Captain Alexandro Malaspina explores the Northwest Coast of North America for Spain. He searches for the elusive Northwest Passage. Because of the population increase from Loyalist migration from the south, the British divide Quebec into two colonies, Upper Canada [now Ontario] and Lower Canada [now Quebec].
1793 Thompson continues his work in Muskrat Country and surveys the waterways between the Nelson and Churchill Rivers.	**1793** Alexander Mackenzie of the NWC reaches the Pacific Ocean at Bella Coola. The revolutionary French government executes King Louis XIV and Queen Marie Antoinette.
1794 The HBC officially recognizes Thompson as a surveyor.	**1794** Jay's Treaty between Britain and the U.S. settles a number of issues lingering from the Revolutionary War period. One of those issues was the continuing British occupation of fur trade forts on American territory. The treaty establishes a neutral boundary commission to settle border disputes. It also restores trade and guarantees Aboriginal Peoples free movement across the border.
1797 Thompson leaves the HBC to join the NWC. This move will provide	**1797** John Adams becomes the second president of the United States.

DAVID THOMPSON AND HIS TIMES

much more opportunity for exploring, surveying, and mapping.

At Grand Portage, the interior headquarters of the NWC on the western shore of Lake Superior, Thompson meets with senior partner William McGillivray and Alexander Mackenzie. His first task is to locate the latitude of fur trading posts west of Lake Superior because some of them might be in future American territory.

1798
Thompson travels through the northern fur trade district. He visits Mandan villages and charts the headwaters of the Mississippi River.

1799
Thompson marries thirteen-year-old Charlotte Small, according to the custom of the country, which lacks clergymen. Charlotte is the mixed-blood daughter of a trader who abandoned his family at the end of his service.

The NWC builds Rocky Mountain House in what is now the province of Alberta.

1800
Thompson's trading and surveying ventures increasingly take him into the Rocky Mountains.

CANADA AND THE WORLD

A group of traders in Montreal, dissatisfied with Simon McTavish's NWC leadership, form a rival XY Company in competition with the NWC.

1798
Napoleon's troops land in Egypt.

British Admiral Nelson defeats the French navy at the Battle of the Nile.

1799
The Dutch East India Company, once a master of global trade, dissolves.

George Washington dies.

Because of disagreements with his partners, Alexander Mackenzie resigns from the NWC. He will join the rival XY group.

1800
North America's first smallpox vaccination is administered in Newfoundland. The HBC will soon provide vaccinations to the native people in the Northwest.

David Thompson and His Times

1801
A daughter named Fanny is born to David Thompson and Charlotte Small at Rocky Mountain House.

1804
A son, Samuel, is born to Charlotte Small and David Thompson at Peace River Forks.

Thompson becomes a partner in the NWC. He works in the Peace River country.

1806
Thompson's attention is directed to exploring the mountains and discovering a transportation route to the Pacific.

Canada and the World

1801
Alexander Mackenzie's *Voyages to the Frozen and Pacific Oceans* is published in London. He will be knighted for his contributions.

1803
The Americans negotiate the purchase of the vast Louisiana territory from the French. The Louisiana Purchase doubles the size of the United States.

The XY Company is now controlled by Alexander Mackenzie.

1804
President Thomas Jefferson sends an expedition led by Meriwether Lewis and William Clark to explore the newly purchased territory and to continue beyond the territory to the Pacific.

Upon the death of Simon McTavish the NWC and XY Company merge.

Napoleon Bonaparte crowns himself Emperor of France in an elaborate ceremony at Notre Dame Cathedral.

1806
Lewis and Clark reach the Pacific Ocean and begin their return journey.

After a Dutch surrender, the Cape Colony in southern Africa becomes a British colony.

David Thompson and His Times

A daughter, Emma, is born to Charlotte Small and David Thompson at Reed Lake House.

1807
Thompson crosses the Rocky Mountains at Howse Pass. His party includes Charlotte Small and their three children.

He establishes a trading post and fort, which he names Kootenay House.

1808
Thompson explores and trades in the region that is now southern British Columbia, Montana, and Idaho.

A son, John, is born to Charlotte Small and David Thompson at Boggy Hall.

1809
Peigan resistance to NWC traders makes life increasingly difficulty for Thompson and others in Kootenay country.

Canada and the World

United Sates Army Lieutenant Zebulon Pike begins his explorations of the American west.

1807
The British Slave Trade Act comes into effect, abolishing slavery within the British Empire.

In the U.S., inventor Robert Fulton's steamboat travels from New York to Albany on the Hudson River and inaugurates the world's first commercial steamboat service.

American President Thomas Jefferson signs the Embargo Act, which closes U.S. ports to all exports and restricts imports from Britain.

1808
The NWC's Simon Fraser descends to the Pacific via the river that will bear his name. Latitude readings reveal that it is not the Columbia River.

John Jacob Astor incorporates the American Fur Company.

Napoleon invades Spain.

1809
Robert Fulton patents the steamboat.

John Molson's steamboat *Accommodation* is launched on the

David Thompson and His Times	**Canada and the World**
	St. Lawrence River. This is the second steamboat in North America.
1810	**1810**
Thompson is on his way to Montreal when he is sent west once more in response to the creation of the American Fur Company by Astor.	John Jacob Astor creates the Pacific Fur Company. He embarks on the *Tonquin* from New York City with thirty-three employees for a six-month journey around the tip of South America. He establishes a trading post at Astoria at the mouth of the Columbia River.
He experiences difficulties with harsh winter weather and rebellious workers. To avoid Peigan interference, he finds Athabasca Pass, which will become the standard route for fur traders.	In Quebec, Governor James Craig suppresses the newspaper *Le Canadien* group for seditious utterances. He seizes the plant and jails the editor Pierre Bédard and others.
He builds trading posts on Pend Oreille Lake and the Flathead River.	
1811	**1811**
In July, Thompson finally reaches the mouth of the Columbia to discover the trading post Fort Astoria with an American flag flying.	At the Battle of Tippecanoe, American forces led by William Henry Harrison in Indiana Territory upset the progress of Shawnee Chief Tecumseh's growing Indian Confederacy.
A son, Joshua, is born to Charlotte Small and David Thompson at Fort Augustus.	
1812	**1812**
Thompson retires from the fur trade to live in the Montreal area. The NWC honours him with an annual payment of £100 plus his full share of the profits for three years. In return he is to prepare maps.	The Napoleonic Wars of Europe spill over into North America with an American Declaration of War against Britain.
	In the final major battle of the year, British general and Governor

David Thompson and His Times

With access to clergy for the first time since their wilderness marriage, David and Charlotte regularize their marriage in the Scotch Presbyterian Church.

1813
A son, Henry, is born to the Thompsons in Terrebonne village, near Montreal.

1814
Thompson completes a large map of the Northwest from Lake Superior to the Pacific. This map will hang for several years in the great hall at Fort William, the NWC interior headquarters on Lake Superior.

The Thompsons experience the deaths of their children John and Emma, two of their five children born while David Thompson was exploring in the West.

1815
At the end of his three years as map-maker Thompson retires from the NWC. He is given the usual seven-year allowance of a one-hundredth share of the company's profits. An increasing trade

Canada and the World

of Upper Canada Sir Isaac Brock is killed at Queenston Heights.

1813
Two British Canadian victories, at the Battles of Chateauguay and Crysler's Farm, prevent American forces from advancing on Montreal.

Americans successfully confront British forces at Moraviantown in Upper Canada, resulting in the death of Chief Tecumseh.

1814
The Treaty of Ghent officially ends the War of 1812 in North America.

1815
In Europe, at the Battle of Waterloo, massed armies led by the Duke of Wellington defeat Napoleon for the final time.

DAVID THOMPSON AND HIS TIMES

war with the HBC, however, will see profits diminish.

The Thompsons move to a farm at Williamstown, Upper Canada.

A daughter, Charlotte, is born to the Thompsons in Williamstown.

CANADA AND THE WORLD

John A. Macdonald, future first prime minister of the new nation of Canada, and Otto von Bismarck, future Prussian/German leader, are born.

1817

Thompson accepts a position as astronomer and surveyor with the International Boundary Commission. Established by the Treaty of Ghent, the Commission is to survey the boundary with the United States as far as the Rocky Mountains.

A daughter, Elizabeth, is born to the Thompsons in Williamstown.

1817

The Rush-Bagot Agreement limits the number of warships of the Great Lakes to eight.

John Palliser, who will lead an 1857 British expedition through the British Northwest to British Columbia to assess its suitability for agricultural settlement, is born.

1819

Thompson undertakes increased responsibilities in managing the field operations of the survey crews of the International Boundary Commission.

A son, William, is born to the Thompsons in Williamstown.

1819

A British cavalry unit charges into a crowd of 60,000 protesters at Peterloo near Manchester, killing eleven and wounding four hundred. In depressed economic conditions following the end of the Napoleonic Wars, these men, women, and children are protesting the high price of bread and calling for an end to import tariffs.

1820

Thompson is appointed a justice of the peace in Glengarry County, Upper Canada.

1820

British King George III dies.

Explorer Sir Alexander Mackenzie dies in Britain.

David Thompson and His Times / **Canada and the World**

1821

After years of disastrous rivalry, the HBC and the NWC merge under the name of the HBC. Except for a small HBC office, the Montreal area loses its historic connection with the fur trade, which has existed since the early years of the French colony in the seventeenth century.

The U.S. takes possession of Florida, which it recently purchased from Spain.

1822

Thompson begins a four-year period surveying from Lake Superior to the Lake of the Woods.

A son, Thomas, is born to the Thompsons in Williamstown.

Brazil declares its independence from Portugal.

1824

Another son, George, is born to the Thompsons in Williamstown, but the baby dies at the age of seven weeks.

The Geological Survey of Canada is founded.

The Lachine Canal bypassing the rapids at Montreal is completed.

1827

A daughter, Mary, is born to the Thompsons in Williamstown.

Sandford Fleming (future Canadian railway surveyor) is born in Scotland.

1829

David and Charlotte's thirteenth and last child, a daughter named Eliza, is born in Williamstown.

Andrew Jackson is inaugurated as seventh president of the U.S.

The first Welland Canal bypassing Niagara Falls is completed.

David Thompson and His Times	Canada and the World
1830 With the loss of fur trade income and the completion of boundary commission work, the Thompsons slide into poverty.	**1830** William IV becomes King of Great Britain and its colonies. Belgium wins its independence from Holland.
1843 The British government pays Thompson $150 for a new version of his map of the Northwest. He completes work on his atlas, which maps the land from Hudson's Bay to the Pacific, covering an area of over 3.9 million square kilometres. Between 1843 and 1850, Thompson works on writing his *Narrative*, a manuscript of his adventures that he hopes to publish as a book.	**1843** In the U.S., the first major wagon train leaves Missouri on the Oregon Trail. The HBC builds Fort Victoria on the tip of Vancouver Island to strengthen the British claim to the region and to guarantee access to the mainland through Puget Sound.
1845 Their resources depleted, David and Charlotte reluctantly move in with a daughter and son-in-law.	**1845** An enlarged Welland Canal between Lakes Erie and Ontario opens. *Scientific American* begins publication. The Irish potato famine begins.
	1846 The forty-ninth parallel is accepted as the boundary between the United States and the British lands to the north. The British retain Vancouver Island.

David Thompson and His Times

1851
Having experienced failing eyesight for some years, Thompson is now totally blind.

1857
David Thompson dies at his daughter's residence in Longueuil on February 10, at the age of 86. His wife Charlotte Small dies later in the same year. Both are buried in unmarked graves.

Canada and the World

1851
The Great Exhibition of the Works of Industry of All Nations, held in the Crystal Palace, opens in Hyde Park, London.

The first YMCA in North America opens in Montreal.

The *New York Times* begins publication.

Reuters news service begins operation.

1857
Bytown, now renamed Ottawa, is selected as the capital of the colonies of the United Canadas.

John McLoughlin, former head of HBC Pacific District, dies.

The British expedition led by John Palliser begins its work in western North America to determine the suitability of the HBC lands for agricultural settlement.

The Grand Trunk Railway is completed from Windsor to Montreal.

In Japan, an earthquake kills over 100,000 people in Tokyo.

1868
The HBC relinquishes its trading monopoly and transfers title of Rupert's Land to Canada in exchange for £300,000, the lands

David Thompson and His Times | **Canada and the World**

around its trading posts, and 1/20th of the good agricultural lands on the Prairies.

In addition to the fur trade, the HBC's business now includes real estate and the selling of consumer goods to the new settlers.

1869
Many residents of the HBC Selkirk settlement in the Red River Valley object to the lack of consultation in the transfer of title to Canada. Led by Louis Riel, the Métis and others rebel and force negotiations.

1870
The Red River Rebellion thwarts Canadian plans to create a colony of the old HBC lands. Rebels demand and receive provincial status for their settlement. The new province of Manitoba is born.

1871
Sandford Fleming accepts the position of Chief Engineer of the proposed Pacific railway.

British Columbia enters Confederation as the sixth Canadian province.

1880
A contract for the construction of a railway from Montreal to the Pacific is awarded to the Canadian Pacific Railway (CPR) syndicate.

David Thompson and His Times

Canada and the World

1885

The CPR is completed.

1914

Joseph Tyrell begins editing David Thompson's unfinished *Narrative* of his fur trade years for publication.

1914

The ocean liner *Empress of Ireland* sinks in the Gulf of St. Lawrence, resulting in the loss of 1024 lives.

The Panama Canal opens, making travel between the Atlantic and Pacific possible without travel around the tip of South America.

The First World War begins.

1916

The Champlain Society publishes *David Thompson's Narrative*.

1916

More than one million soldiers die in the Battle of the Somme in France.

Canadian women in Manitoba receive the right to vote, followed by those in Saskatchewan and Alberta.

Alberta introduces Prohibition of the sale of alcohol.

In Ottawa, the Centre Block of the Parliament Buildings burns.

1922

At the former site of Kootenay House on Lake Windermere a group of Canadians gather to honour David Thompson. Over 300 First Nations people participate. Poet Laureate Bliss Carman and John Murray Gibbon, publicity

1922

A British group enters the tomb of Egyptian King Tutankhamun, the first entry since the tomb was sealed 3,000 years ago.

David Thompson and His Times

director of the CPR, attend, and Bliss Carman writes a poem to commemorate the event.

1927
On the seventieth anniversary of Thompson's death, special ceremonies are held in Montreal to unveil a monument at the site of his unmarked grave in Mount Royal Cemetery.

1957
Canada issues a postage stamp with David Thompson's picture on it.

1962
A new edition of Thompson's *Narrative* edited by Victor Hopwood is published by the Champlain Society. This edition contains additional material and is titled *David Thompson's Narrative, 1784-1812*.

Canada and the World

1927
The first transatlantic telephone call is made from New York City to London.

Charles Lindberg makes the first non-stop solo flight across the Atlantic. He flies from New York to Paris.

The Ford model A replaces the model T.

1957
John Diefenbaker of Prince Albert, Saskatchewan, becomes prime minister of Canada.

The Soviet Union launches *Sputnik 1*, the first artificial satellite to orbit the earth.

1962
The Trans-Canada Highway opens.

With the launch of the communications satellite *Alouette 1*, Canada becomes the third nation to reach space.

The drug thalidomide, which resulted in serious birth defects, is banned in Canada.

DAVID THOMPSON AND HIS TIMES	CANADA AND THE WORLD
1994 Barbara Belyea edits an edition of David Thompson's *Columbia Journals*.	**1994** The HBC donates its company records to the Province of Manitoba. NAFTA, the North American Free Trade Agreement between Canada and the U.S., begins. The Channel Tunnel between England and France opens. Nelson Mandela becomes the first Black president of South Africa.
2005 David Nisbet's illustrated book, *The Map-Maker's Eye: David Thompson on the Columbia Plateau*, is published.	**2005** Saskatchewan and Alberta celebrate their centennials. The undefeated Canadian junior men's hockey team defeats the Russian team to win the World Junior Championship.
2006 The David Thompson North American Bicentennial Partnership plans for commemorations between 2007 and 2011 to honour Thompson's many contributions.	**2006** Jerry Zucker, a U.S. businessman, becomes the majority shareholder and the first American Governor of the HBC.

Acknowledgments

I have strived to be accurate in presenting the historical facts about David Thompson. The places he visited, the dates, the people he met, and his view of events, are all given according to his own written narrative or taken from other historical sources. Thompson revealed scant details about his childhood years. We know he was orphaned and at age seven, sent to the Grey Coat Charity School in London where he received a remarkably good education considering his circumstances. Research into the school and that period in London's history revealed a harsh and unforgiving world, particularly for poor and orphaned children. Some scenes in chapters one and two, which detail his early years, were created from this research. For the remainder of the book, the created scenes and dialogue adhere closely to Thompson's own well-written accounts either taken from *Travels in Western North America 1784-1812*, edited by Victor Hopwood or from Thompson's unpublished journals retrieved from the Archives of Ontario.

The "Travels" were written by Thompson when he was an old man and in the last years of his life. He starts his story with his arrival, in 1784, at the Hudson's Bay Company post at Churchill Factory. He was fourteen years old and fresh from London. He then describes, in modest terms, a magnificent thirty years of epic travels across the continent, drawing from his extensive field notes and from memory. His narrative ends with the news of the outbreak of the War of 1812 and with his final rush to Montreal to evade capture by the Americans. Although the veracity of a few points in Thompson's narrative may be challenged by some historians, I have chosen to follow faithfully the explorer's own words as my guide for retelling his remarkable story.

No book is completed without help. John Wilson, an outstanding writer, generously guided me through the bumpy terrain of the writing and publishing world. My wife Sue faithfully edited the many revisions of this work, and was a constant source of support. Vladimir Konieczny, a fellow writer, offered many constructive comments, as did Jennifer Wilson on a final draft. Rhonda Bailey gave professional editorial guidance. Barbara Belyea and David Malaher provided helpful information on source material. Finally, I am indebted to my daughters Erin, Brianne, and Cindy, and grandson, Thomas, who continue to teach me about the important things.

XYZ Publishing would like to thank Wildsight and the David Thompson Heritage Lands Project for permission to reproduce the painting by Joe Cross of David Thompson crossing the Athabasca Pass in 1811. Wildsight seeks to commemorate David Thompson's explorations of the Canadian Rockies and the Columbia River route to the Pacific and to conserve the wilderness characteristics and outdoor experience of the David Thompson Heritage Lands. Information on the David Thompson Heritage Lands Project is available at http://www.wildsight.ca.

XYZ Publishing would also like to thank Ross MacDonald, David Malaher, and Andreas Korsos of the North American David Thompson Bicentennials Partnership for their assistance. The North American David Thompson Bicentennials will take place on a continent-wide basis between 2007 and 2009 inclusively. For information, visit the website at www.davidthompson200.org/.

Sources Consulted

HOPWOOD, Victor G. (Ed.). *David Thompson's Travels in Western North America, 1784-1812*. Toronto: Macmillan of Canada, 1971.

JENISH, D'Arcy. *Epic Wanderer: David Thompson and the Mapping of the Canadian West*. Toronto: Doubleday, 2003.

JENKINS, Kathleen. *Montreal, Island City of the St. Lawrence*. New York: Doubleday, 1966.

MACKAY, Donald. *The Square Mile: Merchant Princes of Montreal*. Vancouver: Douglas & McIntyre, 1987.

MACKIE, Richard Somerset. *Trading Beyond the Mountains: The British Fur Trade on the Pacific 1793-1843*. Vancouver: UBC Press, 1997.

Malaher, David G. "David Thompson's Surveys of the Missouri/Mississippi Territory in 1797-98." 9th North American Fur Trade Conference & Rupert's Land Colloquium. University of Missouri at St. Louis, Missouri, 2006.

NEWMAN, Peter C. *Company of Adventurers*. Markham: Penguin, 1985.

———. *Caesars of the Wilderness*. Markham: Penguin, 1988.

SOBEL, Dava and ANDREWS, William J.H. *The Illustrated Longitude: The True Story of a Lone Genius Who Solved the Greatest Scientific Problem of His Time*. Toronto: Penguin, 1995.

SEBERT, L.M. "David Thompson's Determination of Longitude in Western Canada." *The Canadian Surveyor*, 35, no. 4: 405-414. March 1981.

SMITH, James K. *David Thompson: Fur trader, Explorer, Geographer*, Toronto: Oxford University Press, 1971.

SMYTH, David. "David Thompson's Instruments and Methods in the Northwest, 1790-1812." *Cartographica*, 18, #4, 1-17. September, 1981.

Thomson, Don W. *Men and Meridians: The History of Surveying and Mapping in Canada, Volume 1, prior to 1867*. Ottawa: Queens Printer and Controller of Stationery, 1966.

TYRELL, J.R. *A Brief Narrative of the Journeys of David Thompson in North-Western America*. Toronto: Copp, Clark & Company, 1888.

———. "David Thompson and the Rocky Mountains." *Canadian Historical Review*, 39-45. March, 1934.

Index

Numbers in *italics* indicate pages of illustrations or photographs.

Printed in November 2006
at Marquis,
Cap-Saint-Ignace (Québec).